D1035191

PAST CLIMATES

PAST CLIMATES

Tree Thermometers, Commodities, and People

BY LEONA MARSHALL LIBBY

Foreword by Rainer Berger

UNIVERSITY OF TEXAS PRESS, AUSTIN

Printed in the United States of America
First Edition, 1983

Requests for permission to reproduce material from this work should be sent to Permissions, University of Texas Press, Box 7819, Austin, Texas 78712.

LIBRARY OF CONGRESS CATALOGING IN PUBLICATION DATA

Libby, Leona Marshall, 1919–
  Past climates.
  Bibliography: p.
  Includes index.
  1. Paleoclimatology. 2. Dendrochronology. I. Title.
QC884.L5 1983 551.6 82-8648
ISBN 0-292-73019-5 AACR2

TO WILLARD FRANK LIBBY, great scientist and great creature. His serious contemplation of every kind of scientific problem, his analytic testing of conclusions, his care in measuring predictions from the analyses, his love and devotion to teaching, his openhearted help to everyone are inspirations for all.

Contents

Foreword

Leona Libby is one of the modern pioneers in climatic research. This field has been gaining more impetus in recent years, as it holds considerable promise of predicting climatic trends in future decades and centuries. Thus this book is of great interest not only to the scientist but also to economists, financial experts, politicians, and the military.

In recent years, tree ring–based temperature data have been collected which go far beyond the records available to historians. These data can be analyzed by Fourier transforms which identify certain periodicities. The longer the base data record, the more accurate the predictions are apt to be. It goes without saying that climatic changes detected by tree rings have been checked against historic records of the same time as much as that is possible today. The correspondence is astonishing. Therefore, there is considerable promise that the basic methodology is correct.

At present weather forecasting is becoming more accurate for periods on the order of days, weeks, and months. Climatic prognoses have also been attempted for very long times of tens of thousands of years. But the intermediate range in the decades and centuries so far has been an enigma. It is here where tree ring thermometry plays its trump cards.

The degree of accuracy and verification of modern decade- and century-long forecasts needs to be checked against the actual behavior of climate in the years to come. This makes long-term research projects and their funding mandatory if they are to succeed. The customary support for a few years is obviously not appropriate for this type of research, which need not be a widespread effort, however. Yet its potential is enormous in assessing worldwide crop yields, water inventory, heating requirements, stockpiling policies, and construction planning as well as political and military prospects.

So far, tree rings that have been dendrochronologically, that is, calendrically, dated reach back about 7 to 8,000 years into the past. Moreover, wood fragments or tree logs dated by radiocarbon have been found to be as old as 20 to 30,000 years. Thus there is every possibility of extending our base record far back into the Upper Pleistocene with an attendant increase in our understanding of climatic changes in the past. Using mathematics, we can translate this record into the most valuable predictions on which to base our national and international long-term policies.

Rainer Berger
Professor of Anthropology, Geography, and Geophysics
Chairman, Archaeology Program, UCLA

Acknowledgments

It is a pleasure to thank Rudolph Black of the United States Defense Advanced Research Projects Agency, who, in May 1971, funded my proposal that "temperature variations in past climates may be evaluated by measuring stable isotope ratios in natural data banks such as tree rings and varve sequences." Thanks go to William Best of the U.S. Air Force Office of the National Science Foundation, who subsequently provided further funds. Certainly success of the measurements was due to Louis Pandolfi. He and I thank our collaborators, Patrick Payton and John Marshall III, who successfully carried out many of the difficult sample preparations, Bernd Becker of the Landwirtschaftliche Hochschule, Universität Hohenheim, Stuttgart, Germany, and V. Giertz-Siebenlist of the Tree Ring Laboratory, University of Munich, who provided authentically counted and dated chronological sequences of tree rings, as did K. Y. Kigoshi of Gakushuin University, Tokyo, Japan, and Henry Michael of the University of Pennsylvania, Philadelphia, Pennsylvania.

We used the measurements of organic carbon and uranium in a sea core made by Paul R. Doose, Emil K. Kalil, and I. R. Kaplan of the University of California at Los Angeles, and we compared our stable isotope measurements with Hans Suess' measurements of radiocarbon in the bristlecone pine ring sequence prepared by Wesley Ferguson, Suess at the University of California at San Diego and Ferguson at the University of Arizona, Tucson.

We thank Lawrence H. Levine for helping with the UCLA PDP-11 computer and for preparing computer programs.

Most of all we thank Willard Frank Libby, who advised and encouraged us and who took time and care to study our results.

Preface

This book describes the new science of tree thermometry, depending on measuring stable isotope ratios in wood, and its power to reveal climatic changes in the past. The history of the discovery of the existence of stable isotopes of the light elements is reviewed, as is its evolution from the study of radioactive isotopes of the heavy elements.

Knowledge of changes in past climates is relevant to understanding how humans have evolved, how climatic change has influenced their way of living and the developing technology by which they defended themselves against climate exigencies. The method of tree thermometry depends on measuring the ratios of the stable isotopes of hydrogen and oxygen in tree rings. These elements derive from rain and snow, which in turn originate as water vapor distilled from the surface of the seas. Variations in sea surface temperature cause variations in the stable isotope ratios in the distilled water vapor and therefore changes in their ratios as they are stored year by year in tree rings. By evaluating their changes in tree rings, we are able to determine climate changes of the past, as far back as tree rings may be secured. By applying the method of Fourier analysis to the stable isotope measurements in trees, we can predict climate changes to be expected in the near future. Such predictions have not yet been tested, but they should certainly be studied for their potential influence on world food and energy supplies and on national defense.

Methods of prediction of future climate changes using computerized models are described and compared with the method of tree thermometers.

The analytic chemistry used in measuring stable isotope ratios in wood is described with care, so that others can use it to measure trees from all over the world. In this way we may learn if climate changes have been cosynchronous in both hemispheres. Measurements of carbon isotope ratios are described, but they are not as useful because carbon does not derive from seawater.

After indications of climate change are described, their interaction with the biosphere and with humanoid and human activities is reviewed, including historic prices of commodities and historic wages.

Human mitigation of climate exigencies by invention of houses and of shoes is discussed.

PAST CLIMATES

Introduction: The Discovery of Isotopes

The method of analysis of variations of past climates by tree thermometry depends on the existence of isotopes of stable elements, specifically of two oxygens of 16 and 18 atomic mass units and two hydrogens of masses 2 and 1. These form six varieties of water molecules (D specifies H-2 and H specifies H-1):

$H^{16}OH$	$H^{18}OH$	$D^{16}OD$
$H^{16}OD$	$H^{18}OD$	$D^{18}OD$

Of these, the last two kinds containing two atoms of D are too rare to be useful for the study of tree thermometers.

Surprisingly, the discovery of isotopes of light stable elements derived from the discovery of isotopes of heavy radioactive elements. The first indications of their existence were observed by Ernest Rutherford, at McGill University, Montreal, in the study of the members of the radioactive decay chains. To define the characteristics of the daughters, he used chemistry and in so doing found that some were so similar chemically that they could not be separated by any chemical operation. This was the case with ionium and thorium. The same is true of mesothorium I, which Otto Hahn in Berlin had discovered in 1907 and found that it cannot be separated from radium; in the same way the final decay products of uranium and thorium cannot be separated from lead (fig. I-1).

Rutherford's collaborator in Montreal, Frederick Soddy, taking a professorship at Glasgow University, brought order and understanding to the puzzle of the chemical similarity of these various radioactive and stable isotopes, incorporating the work of many people who were contributing to the knowledge of isotopes, namely, that elements exist as mixtures of different constituents which are identical in chemical behavior. Chemical analysis classifies according to the external systems of electrons which surround the nuclei, whereas radioactivity, characterized by the kind of particle emitted and the lifetime for emission, depends on the internal structure of the nucleus. Chemical analysis showed that the same electron configuration may surround different interiors in the nucleus. Yet their overall electrical neutrality implies that the different interiors must all have the same electric charge on the nucleus to balance the same electron configuration.

Soddy and Sir William Ramsey showed in 1903, at the University College of London, that helium was being continuously generated from radium in a spectroscopically detectable quantity and that the helium originated as alpha particles, that is, helium nuclei energetic enough to ionize gases, so that radon, the daughter of radium, should weigh 4 mass units less than radium and its nucleus should have 2 less positive charges.

At the time of Soddy's Nobel lecture in 1921, it was known from the chemical analyses of Frédéric Joliot and Marie Curie that radioactivity was restricted to the last two of the then known elements, uranium and thorium, and the radioactive decay chains of these elements. It was further known, by the nature of the ionization in gases and its measurement by sensitive and accurate instruments which had been rapidly developed for the purpose, which daughters in the chains emitted alpha particles, beta particles (electrons), and gamma rays. Except for the uranium and thorium decay chains, the only other radioactivities known were the beta emitters, potassium and rubidium.

Of the many similar, newly discovered radioelements, only radium had been prepared pure, and its visible light spectrum and atomic weight had been determined as well as its radioactive decay heat, 133 calories per gram per hour. For the various forms of radium, thorium, and lead, many people proved the chemical identities in dozens of different ways. In particular, Soddy found it impossible to separate mesothorium from radium by fractional crystallization of chlorides of mesothorium, radium, and barium. Contrary to the suggestion that barium salts entrain or absorb mesothorium in preference to or differently from radium, the radioactive constituents were concentrated by recrystallization over and over again without the slightest observable change in the ratio of their amounts as far as it could be determined by the most careful measurements of their radioactivities (Soddy 1921).

In this way Soddy became convinced that he was dealing with complete chemical identity and that among chemically identical

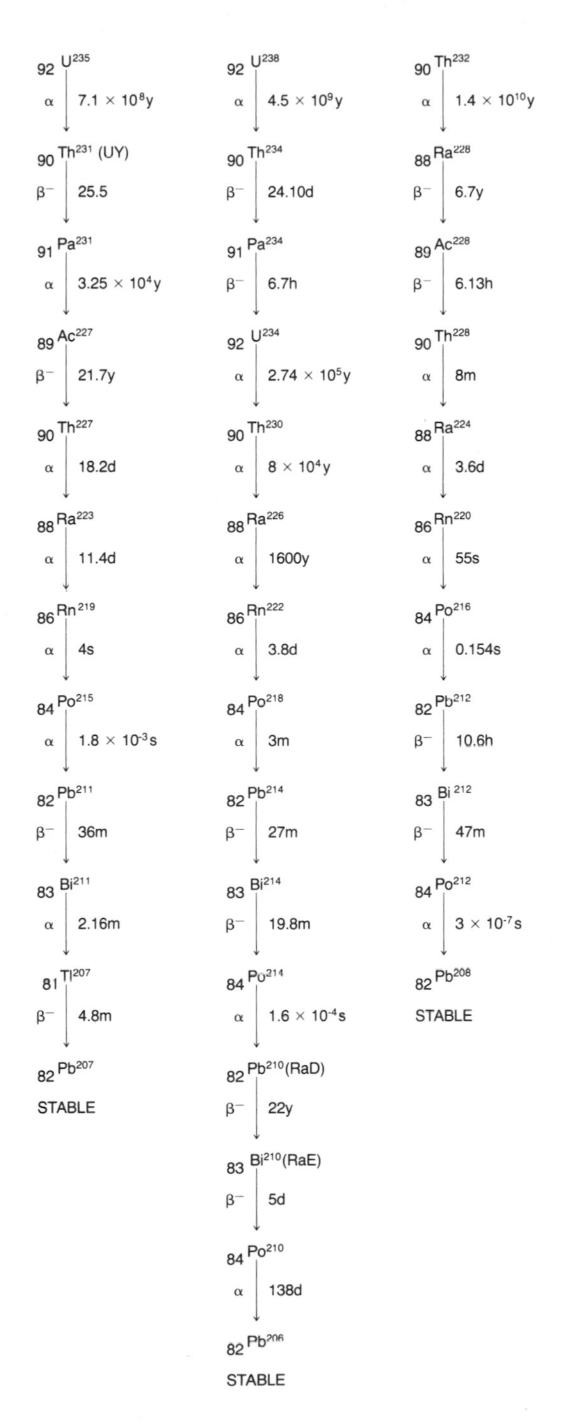

FIG. I-1. Radiodecay schemes for three naturally occurring elements.

radioelements, differences of atomic weights of the parent elements and the number of alpha particles emitted, the atomic weight of ionium (230) and radiothorium (228) must differ by two and four units respectively from that of chemically identical thorium (232). He concluded that, if elements were mixtures of chemical identities differing by whole units of atomic mass, a chemist could never have found it out.

He further conjectured that the recognition that elements of different atomic weights may have identical chemical properties would probably have its most important application among stable elements, where the absence of radioactivity makes it impossible to determine their different masses. Chemical homogeneity was no longer a guarantee that a given element was not a mixture of nuclei of different atomic weights or that a measured atomic weight was not an average of different atomic weights.

Soddy had in mind the knowledge, already ascertained by Theodore Richards (1914), that the atomic weight of radiolead is remarkably lighter than that of ordinary lead. Richards measured them as 206.08 and 207.2 respectively. He reported that the study of other properties of these leads and their salts showed that weight was the only observable difference between the two kinds of lead. Their atomic volume, electrical behavior (plating out from solution by electrolysis), etc., are identical and their optical spectra are very nearly, if not exactly, alike. Richards remarked, "Ordinary lead is shown to possess a constant atomic weight, no matter what its source, provided that uranium minerals are not present in the source." Here in a single sentence is born the postulate of the existence of isotopes.

Richards made his life's work the accurate measurement of the atomic weights of the elements. "If man could only understand more about the elements, further light might develop in all of the complicated phenomena which ultimately depend on their properties. Until we understand the significance of the atomic weights, we cannot hope to understand the atoms themselves nor the countless substances which are made from them. If our ancient universe ever had a beginning, the conditions governing that beginning must be engraved in the atomic weights."

Richards, the first American to be honored by the Nobel Foundation, was educated at Harvard, joined its faculty, and was made professor there in 1901. He was a teacher of Gilbert Lewis, who constructed the physics and chemistry departments at the University of California, teaching among other great scientists Harold Urey. After

Urey's discovery of deuterium, Lewis developed methods to separate heavy water from light water (and fed the first drop to a mouse). So goes the chain of great science from teacher to student.

The next major advance was made by F. W. Aston (1922), who devised a way to prove that isotopes of the elements exist, by a physical method depending on mass and charge, different from the chemical methods used by Soddy and the others; Aston described it as "analysis of 'positive rays.'" He was research assistant to Sir J. J. Thompson in Thompson's work on positive rays, using a device in which atoms were ionized by an electric discharge in a vacuum, then deflected by an electric field in one direction and by a magnetic field in a direction at right angles to the first one. The result was to deflect the ionized atoms so that, when they struck a fluorescent screen or photographic plate, they made a parabolic streak depending in position on the charge to mass ratio of the ions. Thus Thompson had made a primitive mass spectrograph, a very insensitive one. The fact that the streaks were definite parabolas and not mere blurs was the first direct proof that atoms of the same element had approximately the same mass.

In 1912, Thompson put neon into his apparatus and observed a double parabola, the brighter component corresponding to neon-20 and the fainter component to neon-22, in agreement with the atomic weight of 20.20 measured for neon by Richards if there were 10 percent of neon-22. Thompson was of the opinion that the fainter component could not be caused by any compound in neon gas because it is chemically inert; the faint line represented a previously unknown constituent of neon.

As Aston said, "The suggestion that atoms of different weight could exist in the same element had just been promulgated and the facts could be explained very well by the suggestion that neon was a mixture of two such bodies. I therefore undertook to investigate this point as fully as possible."

First Aston used a method of diffusion of neon through clay. After several months of diffusion, he obtained a small but definitive difference of 0.7 percent between the heaviest and lightest fractions of the gas, which fitted well with predictions made using Rayleigh's diffusion theory, pointing to the conclusion that neon is a mixture of isotopes.

His researches were interrupted by World War I; when he resumed them in 1919, the existence of isotopes in the radioactive decay series and in their final decay product, lead, had been verified. It was generally agreed that isotopes exist. Realizing that isotope separation by diffusion was exceedingly tedious and could be only partial at best with his equipment and methods, he returned to separation of neon by physical methods. He made a mathematical examination of an ionized beam by crossed electric and magnetic fields, and he introduced metal slits to sharpen up the beam dimensions. In Thompson's apparatus, the beam had been defined by passing through a tube.

In making the theoretical and engineering analyses, Aston was "fortunate enough to hit on the focusing principle for the instrument. Since it is a close analog of the ordinary optical spectrograph and gives a spectrum depending on mass, the instrument is called a mass spectrograph." When neon was introduced into his apparatus, lines appeared at masses 20 and 22 relative to oxygen at mass 16. "Ten percent of neon 22 would bring the mean atomic weight to the accepted value of 20.20, and the relative intensity of the lines agrees well with this proportion. The isotopic composition of neon was therefore settled beyond a doubt."

Next Aston discovered the two isotopes of chlorine of masses 35 and 37 but found no sign of a line at 35.46, the atomic weight of chlorine gas determined by Richards. Similarly, he found potassium 40 and 36, the latter at about 3 percent abundance, sufficient to account for the mean atomic weight determined by density measurements; he also measured the atomic weight of many other elements.

In particular, he measured helium at mass 4. For the hydrogen molecule, H_2, he measured a mass of 2.012 to 2.018, corresponding to a mass of 1.008 for the hydrogen atom (his mass spectrometer could not focus such a small mass as hydrogen with only one charge on it). Taking into account the difference in mass between the helium atom, assuming it to be formed from four hydrogen atoms, and the mass of four hydrogen atoms, he computed that if 1 gram of hydrogen were transformed to helium the energy liberated would correspond to 200,000 kilowatt-hours. He concluded, "We have here at last a source of energy sufficient to account for the heat of the sun." Albert Einstein received the Nobel Prize in the same postwar session in Stockholm as did Aston, for special relativity and the postulate that $E = mc^2$ which Aston used in making his calculation; namely, mass m could be converted to energy E and vice versa.

With a later, much improved spectrograph, Aston arrived at a mass for hydrogen of 1.00778 relative to oxygen-16. The most accurate determination of the atomic weight of natural hydrogen relative to natural oxygen by chemical methods gave the value 1.00777. The agreement was excellent.

But, in 1929, W. F. Giauque and H. L. Johnston found in studying the emission band spectra of molecular oxygen that it contains atoms not only of mass 16 but also of mass 18 to 1 part in 630, as well as a smaller admixture of mass 17. This meant that the scale of chemical atomic weights, in which the weight of the mixed element oxygen is set equal to 16, is not the same as Aston's mass scale in which the atomic weight of the lightest oxygen isotope is set equal to 16. (Aston's mass spectrograph was still too insensitive to allow him to observe the lines of oxygen at masses 18 and 17.)

On this account, Aston's value for the atomic weight of hydrogen, 1.00778, had to be corrected to 1.00756 in order to compare it with the chemical atomic weight of hydrogen, 1.00777. The difference seemed too great to be accounted for by experimental errors. In 1931, R. T. Birge and D. H. Menzel proposed that natural hydrogen contains atoms of mass 2 in addition to those of mass 1 and that the concentration of the former is about 1 in 4,500.

The neutron had long since been discovered, and atomic nuclei, once thought to consist of protons and electrons, were now known to be made of protons and neutrons, so that a hydrogen atom of mass 2 made of a proton and neutron and electron was plausible and there was an empty place it could occupy in the isotope table, as was true also for a hydrogen isotope of mass 3, which Urey referred to as "tritium."

Harold Urey determined to search for H-2 (he named H-1 "protium" and H-2 "deuterium"), but it would be so rare that it would have to be concentrated in some way, for no isotope so low in abundance could have certainly been detected by the methods known at the time. He concentrated deuterium by distillation of liquid hydrogen in order to facilitate its detection by measuring the wavelengths of its emitted Balmer series line spectra relative to those of H-1 (Urey 1935). From the thermodynamic properties predicted by computation, Urey calculated the expected vapor pressures of the molecules H_2, HD, and HT (where D is H-2 and T is H-3; see equation 1 in appendix 1).

For the ratio of the vapor pressures, use of this simple theory gives $P(H_2) / P(HD) = 2.23$, and $P(H_2) / P(HT) = 3.35$, showing that a very effective concentration of deuterium should be obtained by distillation of solid hydrogen at its triple point. However, as Urey points out, "Of course it was impossible to be certain that these differences would apply to the liquid state, but it was a reasonable postulate that at least some of the effect would persist into the liquid state."

To evaluate the ability of the spectrograph available to him to detect the rare deuterium isotope after concentration, Urey calculated the emitted wavelengths of the Balmer series from H_2 and those expected from HD using the Bohr theory. He calculated for the Balmer alpha lines a wavelength difference of 1.787 angstroms and for the beta lines a difference of 1.323 angstroms, well within the capability of the resolution of their spectrograph, a concave grating having 15,000 lines per inch on a 21-foot circle.

F. G. Brickwedde of the United States Bureau of Standards prepared liquid hydrogen and evaporated 4 liters, held near the triple point, to approximately 1 cubic centimeter. Urey and his assistant, G. M. Murphy, put the sample into a discharge tube, photographed the Balmer emissions, and found the strong and weak lines of H and D in both the alpha and the beta pairs to differ by 1.79 angstroms and 1.33 angstroms respectively. They had proved that deuterium exists in natural hydrogen. No evidence was seen for a hydrogen isotope of mass 3, tritium. We now know, of course, that this isotope is being continually produced by cosmic rays in the atmosphere and is present in rainwater in an abundance of a few atoms in 10^{18} molecules, but it does not accumulate because it decays with a radioactive half-life of 12.3 years.

Urey concluded, "Though the deuterium line is easily detectable in natural hydrogen, it would have been very difficult to have definitely established its existence if the concentrated samples prepared by distillation had not been used; for irregular 'ghosts' of a ruled grating might conceivably have accounted for the observed additional lines. Thus the method of concentration was important in proving the existence of this isotope."

Next several researchers demonstrated that differences exist in vapor pressures of molecules containing different isotopes: in neon fractionally distilled into two components at −248.4° C; for H_2 and HD separated by fractional distillation and by electrolysis; and for $H_2{}^{16}O$, $D_2{}^{16}O$, and $H_2{}^{18}O$. Urey systematically computed the isotope fractionations in reactions of diatomic and triatomic molecules, such as:

$$AD + BH \rightarrow AH + BD$$

$$H_2{}^{16}O + HD^{18}O \rightarrow H_2{}^{18}O + HD^{16}O$$

assuming thermodynamic equilibrium; for, as he noted, fractionation effects related to kinetics of chemical reactions are largely unknown, "except in-so-far as they derive from properties directly related by simple kinetic theory to molecular weights as in diffusion processes or to molecular weights and forces between molecules as in the thermal diffusion method of separating isotopes, where the emphasis has largely been on the separation problem rather than on the fundamental properties of the substances" (1947).

Two more isotopes of importance to the biosphere and its history remained to be discovered, namely, hydrogen of mass 3 (tritium) and carbon of mass 14, both radioactive. In the thirties, after Ernest Lawrence's second cyclotron had been built and was running well, Willard F. Libby and his first graduate student, Sam Rubin, set out to discover carbon-14 because the known radioactive isotope of carbon, C-11, with a half-life of 20 minutes, is too short-lived to be useful as a chemical tracer in life processes (all of which depend on carbon). This means that, once it is made (in this case at the cyclotron) and an experiment is begun using it in living tissue, the count rate caused by its disintegration decreases by a factor of 2 every 20 minutes, as measured with counters, so that after 3 hours the count rate in biological systems has decreased by a factor of 500, rendering further measurement difficult.

Some indications that carbon-14 existed were known. In 1936 William Harkins at the University of Chicago and his students, Franz Kurie and Martin Kamen, using a feeble, few millicurie mesothorium-beryllium neutron source together with a cloud chamber filled with nitrogen (alpha particles emitted by mesothorium strike powdered beryllium, causing neutrons to be knocked out of the beryllium nuclei; the chamber contains nitrogen gas which emits energetic protons when struck by neutrons—as the protons fly away, methyl alcohol condenses along their tracks on electrons torn off the nitrogen molecules, and the tracks thus become visible and can be photographed), had shown that neutron irradiation of nitrogen caused proton emission, leaving a carbon-14 atom behind in all probability. The proton tracks showed clearly in the cloud chamber; the existence of the remaining carbon-14 was inferred.

It took Harkins and his students 3 years to photograph and analyze a few hundred proton recoils, requiring about 10 hours per track for each analysis. Carbon-14 was conjectured to be unstable because the known stable isotopes of carbon were already shown to have masses 12 and 13, and its lifetime was guessed to be long compared with irradiation times of minutes or hours with neutrons.

Libby and Rubin planned an experiment in which they would

irradiate about 100 pounds of ammonium nitrate for a few months with neutrons scattered from the University of California cyclotron target. Using their sensitive Geiger-Müller counters, they would be able to detect the decay of carbon-14 if its lifetime were not longer than a few months. The decay rate is equal to the number of radioactive atoms created divided by the lifetime, so the longer the lifetime, the lower the decay rate and the harder it is to detect.

Using the energy of the protons emitted in the capture of neutrons on nitrogen as measured by Harkins, Kurie, and Kamen, Libby and Rubin estimated the energy for C-14 to decay back to nitrogen with the emission of an electron and predicted that it would be about 170,000 electron volts, about the same as that observed for the decay of sulfur-35, and the lifetime would be about 3 months like that of sulfur-35. But this prediction was not correct, for a reason we still do not know today. Instead, the half-life of C-14 was later found to be 5,730 years; so C-14 was not discovered by Libby and Rubin.

Libby says, "Our calculation about the decay energy of C-14 was correct, but its half-life is 5,730 years instead of 3 months. Why this is so is still a mystery, but it is a lucky thing for radio-carbon dating, because a 3 month half-life would be of no archaeological use. So with only a 3 month irradiation of our sample of ammonium nitrate we didn't make enough C-14 to detect even with our sensitive counters." A similar estimate of a few days' half-life was made by the theoreticians J. Robert Oppenheimer and Philip Morrison.

Martin Kamen, after finishing his doctoral researches with Harkins at the University of Chicago, was hired by Ernest Lawrence at Berkeley, whose cyclotron of pole diameter 37 inches was increasingly in use and whose next cyclotron, with a 60-inch pole diameter, was coming into operation. Lawrence was searching for funds to support the operation and maintenance of these machines, funds from, in particular, biologically and medically oriented foundations. He directed that both machines should be diverted to full-time efforts to determine whether long-lived radioisotopes of the biologically important elements hydrogen, carbon, nitrogen, oxygen, and sulfur existed, such as might be useful in biological researches. He appointed Kamen as the resident chemist of the cyclotron staff in charge of this search (Kamen 1963).

In the quest for C-14, Rubin and Kamen tried bombarding beryllium, boron, nitrogen, and carbon with protons, alpha particles, neutrons, and deuterons. After a lot of failures, the first winner was this reaction:

carbon-13 + deuteron → carbon-14 + proton

and the second winner was

nitrogen-14 + neutron → carbon-14 + proton

They burned the irradiated material to carbon dioxide and precipitated the gas as $CaCO_3$ over and over again to prove that the activity they were measuring really was an isotope of carbon, and they used the screen-wall counter designed and built by Willard Libby to count the very weakly energetic radiation. Sam Rubin, in his doctoral work with Libby, had been intensively trained in the vagaries and distempers of the low-level screen-wall counter; without it, he and Kamen would never have been able to detect the weak radiation of C-14.

Like the inference of the existence of C-14 from particle bombardment experiments, the existence of tritium was inferred. Lord Rutherford at the Cavendish Laboratories, Cambridge University, accelerated deuterons in his Cockroft-Walton machine to bombard a target of heavy water. In this way he inferred two new products from the reactions: a mass 3 isotope of hydrogen, which is tritium, and a mass 3 isotope of helium. (Ordinary hydrogen has mass 1 and ordinary helium has mass 4.) The reactions were observed by detection of the neutrons and protons they produced, as follows:

$$D + D \rightarrow {}^{3}He + \text{neutron}$$

$$D + D \rightarrow {}^{3}H + \text{proton}$$

Of these two isotopes, one must be stable and one radioactive according to the empirical rule that, for nuclei of the same atomic mass, the slightly more massive one is radioactive. The masses of ^{3}He and ^{3}H were not then known, so it could only be guessed which was slightly heavier and therefore radioactive. Lord Rutherford guessed that tritium was stable and therefore would be concentrated by electrolysis of water in the commercial production of heavy water by Norsk-Hydro at Rjukan, Norway, because, being stable, it would have been present since time immemorial.

In 1936, Rutherford persuaded Norsk-Hydro to electrolyze a large amount of 99.9 percent heavy water, D_2O, in the belief that stable tritium would thereby be concentrated as DTO in the remaining water. He sent the resulting concentrated sample to Walter Bleakney at Princeton University, a student of F. W. Aston, to look for tritium by mass spectroscopy. Bleakney found that the sample contained no tritium to less than 2 parts in 100,000. This was to be expected if tritium were radioactive and had decayed; but, since Rutherford and Aston expected it to be stable, they interpreted this result as setting an upper limit on the abundance of tritium in nature.

In fact, Hans Bethe and Robert Bacher at Cornell University, in confirmation of this assumption, published a calculation proving that helium-3 was radioactive and tritium was stable. The Rutherford laboratory at Cambridge University was well equipped with Geiger counters which could have been used to show that the heavy water sample, concentrated by electrolysis, was radioactive with tritium, but such is the influence of accepted ideas that Rutherford did not test his sample for radioactivity. He died shortly after, quite unexpectedly, or else in time he might have brought his sample to a Geiger counter.

Luis Alvarez, in Berkeley at the Lawrence cyclotrons, realized that the failure of the mass spectrograph measurement to find tritium in the Norsk-Hydro electrolyzed water sample meant that tritium was the radioactive species and that helium-3 was stable. He had been a student of Arthur Compton, Nobel Laureate for the discovery of elastic scattering of gamma rays from electrons, at the University of Chicago and had joined the Lawrence cyclotron team afterward. He did not have a mass spectrograph, but he knew that a cyclotron is a super mass spectrograph, in which an ion is hit thousands of times by an electronic ping-pong paddle instead of only once in a mass spectrograph. By feeding helium gas from oil wells into the vacuum space between the magnetic poles of the cyclotron, he accelerated helium ions and moved ion detectors to the position where helium-3 ions would circle in the magnetic field; in this way he detected helium-3, thus proving that stable helium-3 exists to 1 part per million in ordinary helium. Thus was finished the discovery of the stable and radioactive isotopes important to biology and the history of biogeophysics.

References

Aston, F. W., 1922, *Les Prix Nobel en 1919–1922*, P. A. Nörstedt & Soner, Stockholm.

Bien, G., and H. E. Suess, 1967, *Symposium on Radioactive Dating and Methods of Low Level Counting*, U.N. Doc. SM 87/55, International Atomic Energy Agency, Vienna.

Birge, R. T., and D. H. Menzel, 1931, Physical Rev. *37*, 1669.

Bottinga, Y., 1967, J. Phys. Chem. *72*, 800.

Craig, H., 1957, Tellus *9*, 1.

Degans, E. T., R. R. L. Guillard, W. M. Sackett, and J. H. Hellebust, 1968, Deep Sea Res. *15*, 1–9.
Herzberg, G., 1945, *Infrared and Raman Spectra of Polyatomic Molecules*, Van Nostrand & Reinhold, New York.
Kamen, M. D., 1963, Science *140*, 584–590.
Li, Y. H., and T. F. Tsue, 1971, J. Geophys. Res. *76*, 4203.
Libby, L. M., 1972, J. Geophys. Res. *77*, 4310–4317.
Mayer, J. E., and M. G. Mayer, 1940, *Statistical Mechanics*, John Wiley & Sons, New York.
Merlivat, L., R. Botter, and G. Nief, 1963, J. Chem. Phys. *60*, 56.
Richards, T. W., 1914, *Les Prix Nobel en 1914–1918*, P. A. Nörstedt & Soner, Stockholm. (His equipment was a laboratory balance.)
Sackett, W. M., W. R. Eckelmann, M. L. Bender, and A. W. H. Be, 1965, Science *148*, 235–237.
Schiegl, W. E., 1972, Science *175*, 512–513.
Soddy, F., 1921, *Les Prix Nobel en 1919–1921*, P. A. Nörstedt & Soner, Stockholm.
Suess, H. E., 1970, J. Geophys. Res. *75*, 2363.
Urey, H. C., 1935, *Les Prix Nobel en 1935*, P. A. Nörstedt & Soner, Stockholm.
Urey, H. C., 1947, J. Chem. Soc. *1*, 562–581.

1. Principles

The onset of testing of hydrogen bombs in the atmosphere led to an understanding of isotope fractionation in the water vapor that distills from the ocean surfaces throughout the world. This information derived from the establishment of a global network of 155 collecting stations in sixty-five countries, beginning in 1953 and continuing to the present, by the International Atomic Energy Agency and the World Meteorological Organization. Monthly meteorological data (amount of precipitation and temperature) were reported, and monthly samples of precipitation were measured for tritium, deuterium to hydrogen ratio, and oxygen-18 to oxygen-16 ratio (IAEA 1953–1971).

The measured ratios are expressed as δ_D and δ_{18}:

$$\delta_D = [((D/H)_s - (D/H)_{std}) / (D/H)_{std}] \times 10^3 \text{ ppt} \quad (1)$$

$$\delta_{18} = [((^{18}O/^{16}O)_s - (^{18}O/^{16}O)_{std}) / (^{18}O/^{16}O)_{std}] \times 10^3 \text{ ppt} \quad (2)$$

where subscript *s* refers to the sample, subscript *std* refers to standard mean ocean water (SMOW), and ppt means parts per thousand.

The error of measurement of δ_D is about 2 ppt; of δ_{18} it is about 0.2 ppt. The plot of world data from 1953 to 1963, of δ_D versus δ_{18} (fig. 1-1), shows that all the measurements for terrestrial surface waters lie on a line with a slope of 8 characterizing Rayleigh distillation of water vapor from the sea surface to form atmospheric precipitation. This plot with its slope of 8 was originally demonstrated by Harmon Craig (1961, 1963) and by W. Dansgaard (1964). The line is expressed by the relation $\delta_D = 8\delta_{18}$ + constant, where the slope of 8 can readily be computed from the measured temperature coefficients for $((D/H)_{liquid} / (D/H)_{vapor})$ and for $((^{18}O/^{16}O)_{liquid} / (^{18}O/^{16}O)_{vapor})$ (Stewart and Friedman 1975).

Rayleigh distillation, a process in which the condensate is immediately removed from the vapor after formation (by fallout of rain and snow in the meteorological case), leads to a higher fractionation than processes which occur at equilibrium, due to kinetic effects which are not theoretically understood, so that it has not yet been possible to compute the value 8 from first principles. To each point on the line of figure 1-1 there corresponds a temperature of distillation (figs. 1-2 and 1-3).

In figure 1-1, the points at very large isotope depletions ($\delta_D \cong -300$ ppt and $\delta_{18} \cong -40$ ppt) have been measured in very cold ice from the bottom of the Antarctic ice cap, laid down in ice ages. Points at small depletions ($\delta_D \cong 0$ ppt and $\delta_{18} \cong 0$ ppt) have been

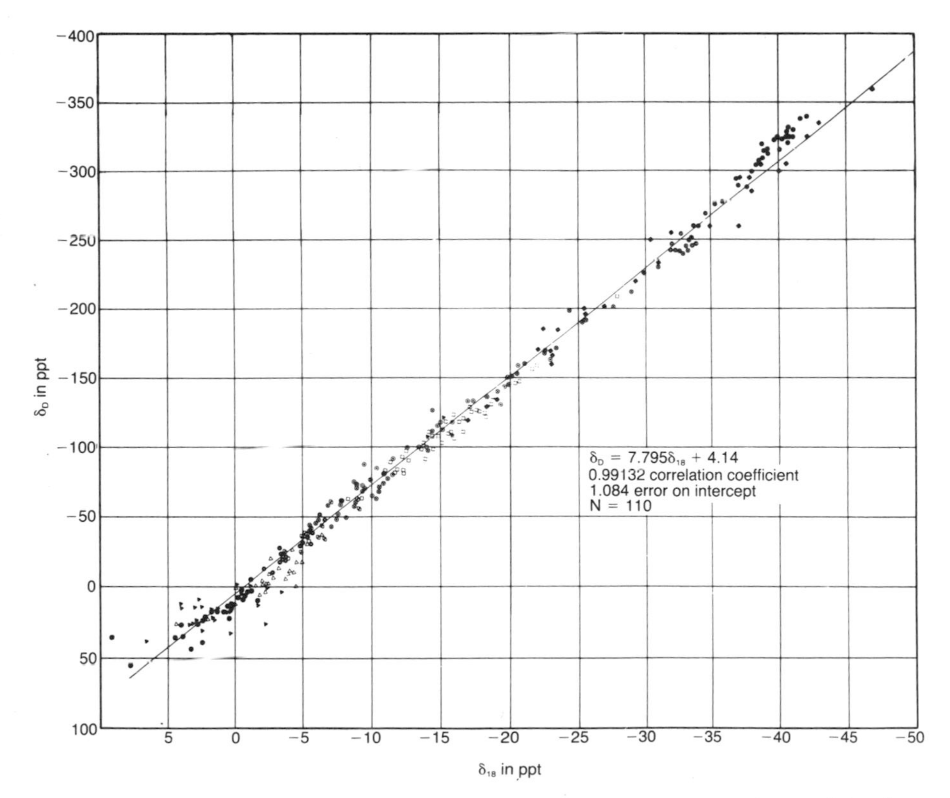

FIG. 1-1. Deuterium isotope ratio versus oxygen isotope ratio for worldwide precipitation (IAEA data, 1953–1963), showing the phenomenological slope of 8.

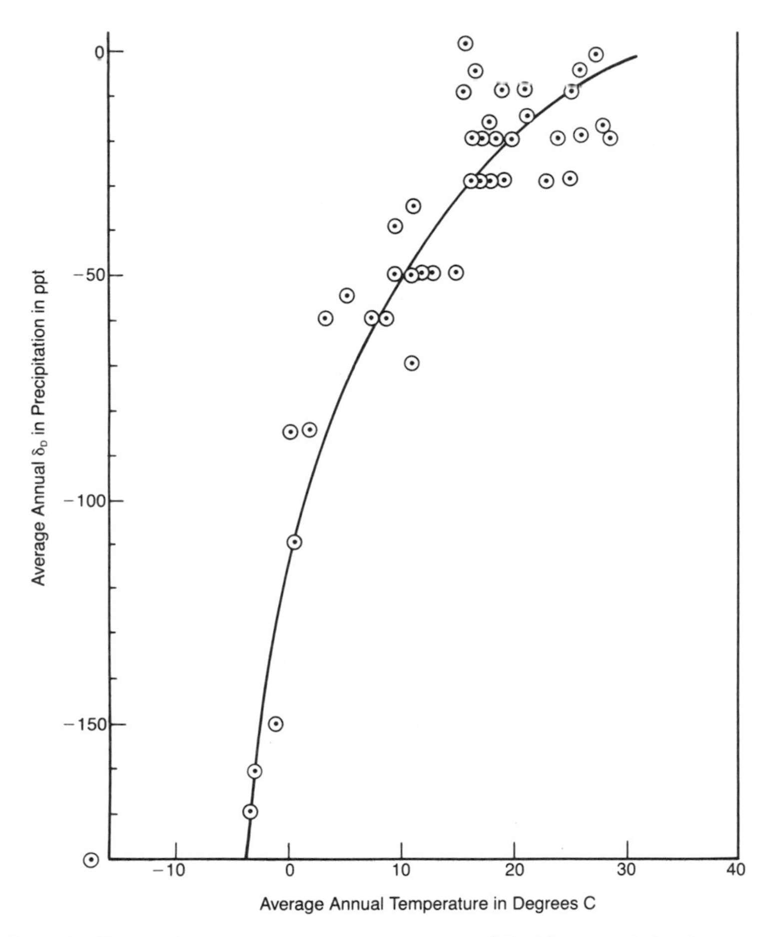

FIG. 1-2. Deuterium isotope ratios in worldwide precipitation versus monthly average air temperatures, showing that for every point on the line in figure 1-1 there is a corresponding average air temperature. The drawn curve is a phenomenological fit to the data.

measured in tropical precipitation distilled from warm oceans. Points between have been measured in middle latitudes.

The IAEA monthly measurements show seasonal variations in that the heavy isotopes are depleted in precipitation when water vapor distills off cold oceans in the winters and enriched in precipitation when water vapor distills off warm oceans in the summers. See figure 1-4 for monthly isotope variations in precipitation, for example, in Stuttgart.

This effect was found in the successive seasonal layers of ice of both the Greenland ice cap and the Antarctic ice cap, showing variations like those in precipitation in temperate regions but, on the average, more depleted in the heavier isotopes, corresponding to the colder average sea surface and air temperature at the poles.

Moreover, in the large scale, in the great depths of the ice cap containing ice laid down 10,000 years ago and more in the last ice age, the ice is more depleted in the heavy isotopes than can be found in any modern-day precipitation. Thus it becomes evident that in the polar ice caps there is stored the history of the surface temperatures of the far northern and far southern oceans, from which dis-

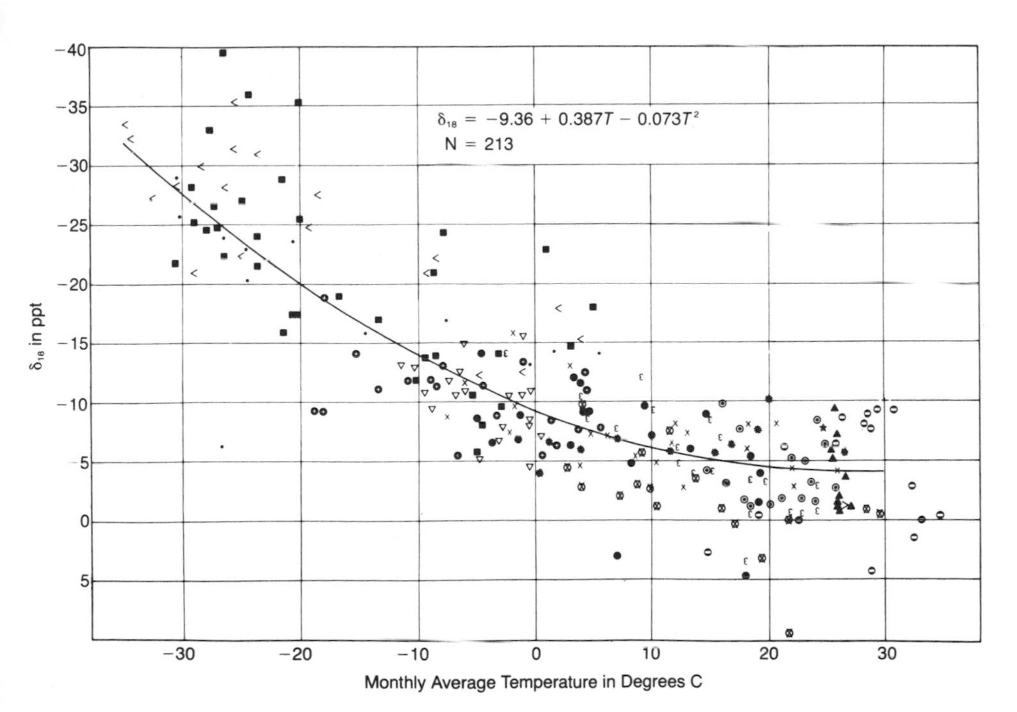

FIG. 1-3. Oxygen isotope ratios in worldwide precipitation (IAEA data, 1966–1967) versus monthly average air temperatures, showing that for every point on the line in figure 1-1 there is a corresponding average air temperature. The drawn curve is a phenomenological fit to the data. The various sorts of data points refer to specific latitudes and longitudes. The average air temperature depends on and varies with the average sea surface temperature from which the precipitation was distilled and was carried to the land by its corresponding air pocket.

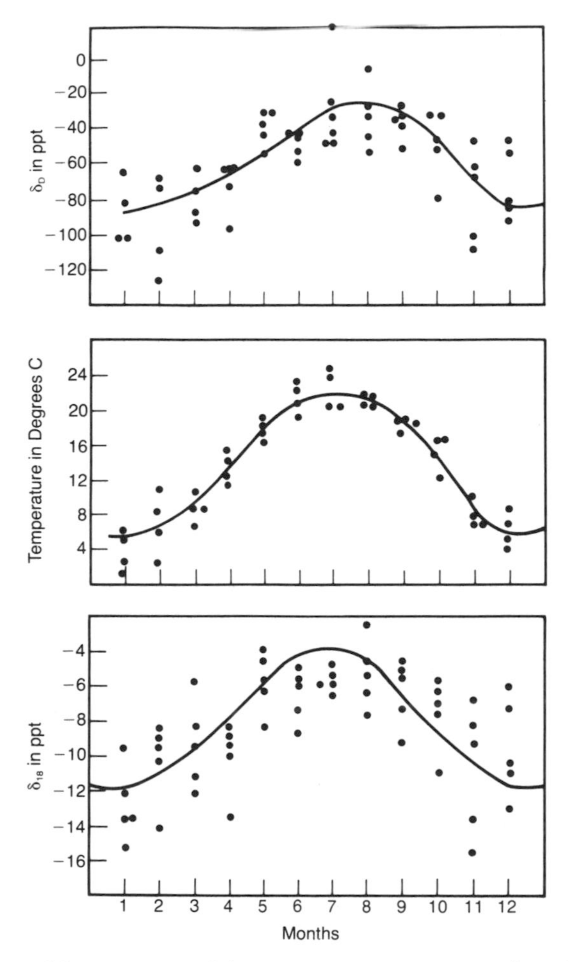

FIG. 1-4. Monthly oxygen and deuterium isotope ratios plotted versus average temperature in Stuttgart precipitation. The drawn curves are phenomenological fits to the measurements.

tilled, for the most part, the historic precipitation laid down in the ice caps.

For temperate regions, the history of the surface temperatures of the oceans is stored in the glaciers of those regions, but glaciers have random advances and retreats which spoil the orderly sequence of the historic yearly ice layers. However, the history of the surface temperatures of the temperate oceans should be stored in the rings of trees which grew in the temperate regions of the world and which subsisted on precipitation distilled from those oceans. Each tree ring should contain some kind of average annual value of the isotope ratios in the precipitation of the year corresponding to the ring.

The wood in each ring is formed according to this reaction:

$$CO_2 + H_2O \longrightarrow \text{wood} + \text{oxygen gas}$$

As for the carbon isotope ratios in tree rings, these derive from and reflect carbon isotope ratios in atmospheric carbon dioxide. There is some evidence suggesting that the ratio $^{13}C/^{12}C$ in atmospheric carbon dioxide varies seasonally in such a way that the isotope ratio is large in the summer. For example, figure 1-5 shows monthly variations in the stable carbon dioxide isotope ratio in atmospheric carbon dioxide at Spitsbergen in the Arctic Ocean, on the Pacific coast of the United States, in Sweden, and at Bariloche, Argentina (Keeling 1960; Ergin, Harkness, and Walton 1970; Olsson and Stenberg 1967; Olsson and Klasson 1970). C. D. Keeling interprets these seasonal isotopic variations as caused by trees preferentially removing $C^{12}O_2$ from the atmosphere in summer when they are growing but not in winter when they are dormant.

Wood is composed approximately of cellulose and lignin. Cellulose is a multiple alcohol of schematic formula $(H\text{-}C\text{-}O\text{-}H)_n$ so that the reaction for the formation of cellulose may be written:

$$CO_2 + H_2O \longrightarrow (H\text{-}C\text{-}O\text{-}H)_n + O_2$$

Lignin contains interconnected aromatic and aliphatic rings (Gould 1966) (fig. 1-6) and aliphatic chains containing about 30 percent oxygen by weight in the form of ether, carbonyl, and hydroxyl bonds. Wood is approximately 25 percent lignin (see ibid.: vii). Its percentage varies somewhat from spring wood to summer wood, and it is possible that its percentage may vary somewhat from ring to ring, so that in principle its variation might affect the temperature coefficient of wood formation.

Assuming the principle of thermodynamic equilibrium to hold in the formation of wood, a very slow process, we have estimated what effect as much as 10 percent variation in the percentage of lignin may have on the temperature coefficient for the formation of wood. We find it to be about 1.5 percent. Since we have felt that a 1.5 percent uncertainty is tolerable within the limits of other errors inherent in the method, we have always analyzed whole wood in our study of isotope variations in tree ring sequences.

In analyzing whole wood, one is confronted by the question of whether to use wet or dry chemistries. Of course, if one decides to

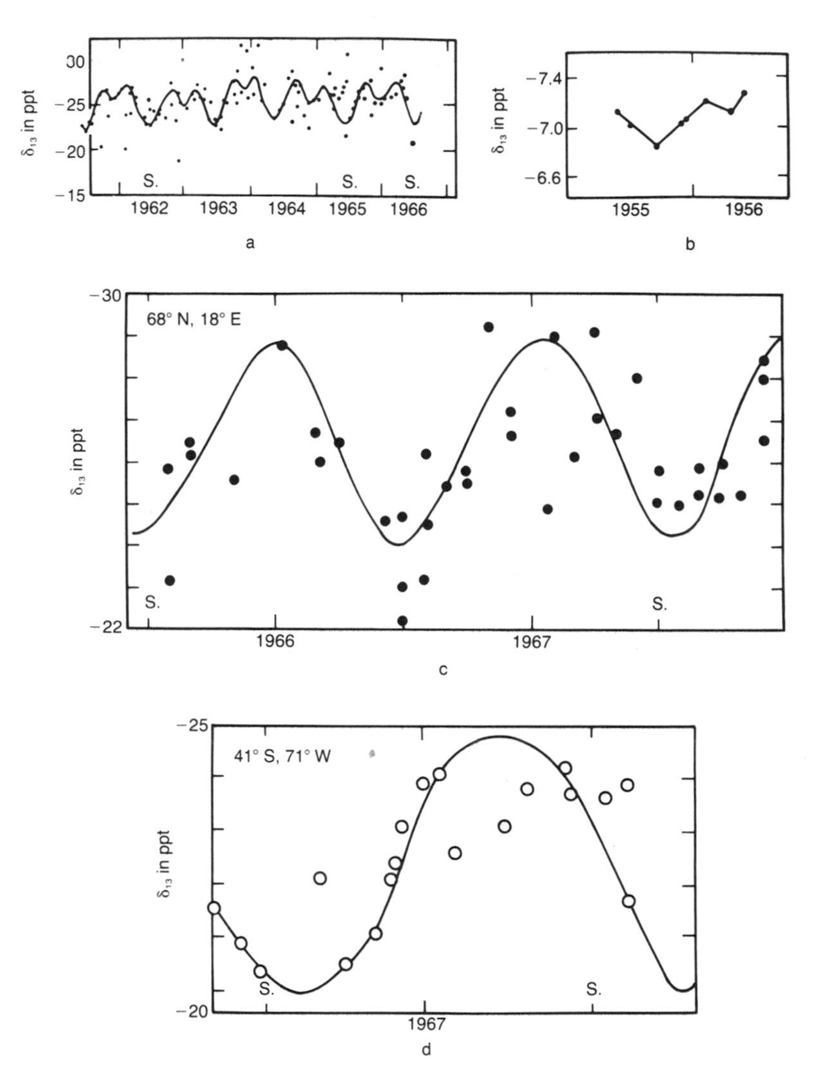

FIG. 1-5. Monthly variations in the stable carbon dioxide isotope ratio in atmospheric carbon dioxide (a) at Spitsbergen, (b) on the Pacific coast, (c) in Sweden, and (d) at Bariloche. The drawn curves are phenomenological fits to the measurements (after Vogel and Lerman 1969; Ergin, Harkness, and Walton 1970; Olsson and Stenberg 1967; Keeling 1960). The symbol δ_{13} is defined for the ratio of carbon-13 to carbon-12 in the same way as δ_D for the ratio of deuterium to hydrogen and in the same way as δ_{18} for the ratio of oxygen-18 to oxygen-16.

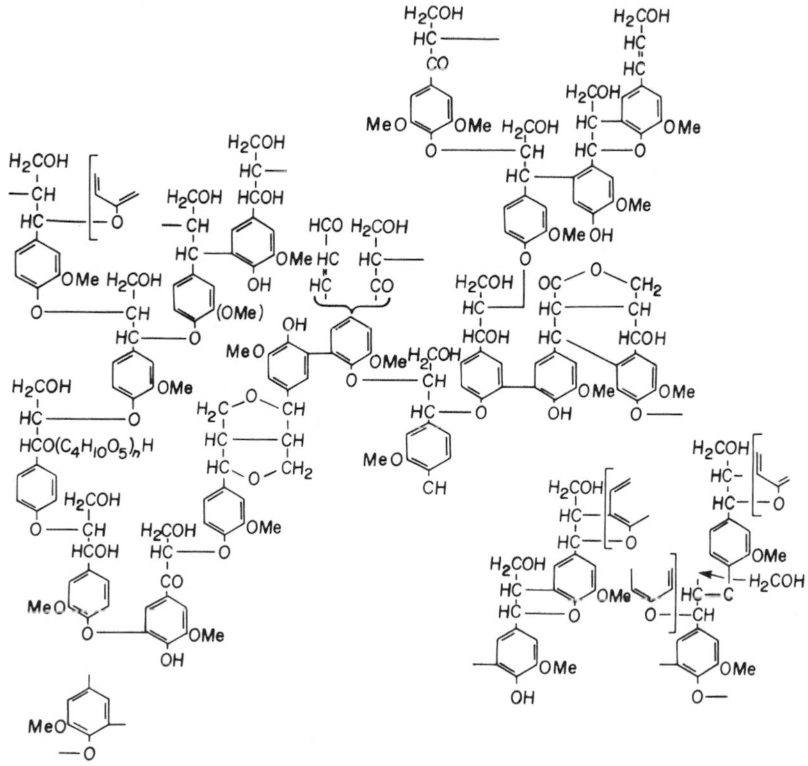

FIG. 1-6. Schematic molecular structure of lignin (after Gould 1966).

separate cellulose from lignin, one is forced to use wet chemistries. It is only in whole wood analysis that dry chemistry becomes possible. With wet chemistries, performed necessarily with hydrogen- and oxygen-containing solvents, there is always the risk of isotope exchange with the solvent. See, for example, the review article of H. Taube (1956). He shows that there exists intimate exchange of hydroxyl oxygen (-O-H) with carbonyl oxygen (-CO-OH) under all conditions of acidity and alkalinity in such liquids as water, ketones, aldehydes, and alcohols. O. Sepall and S. G. Mason (1961) describe the exchange of cellulose and whole wood hydrogens with hydrogen in water as being rapid and effective, leading one to expect similar exchanges with other solvents containing OH groups. The exchange of hydrogen in cellulose with water was found to be 50 percent in

1 hour at 25° C. With dry chemistry there is no possibility for isotope exchange with reagents, and for this reason we have always used dry chemistry.

The reader should be aware of the complexity of the scientific problem and the diversity of expert opinion, as was manifested in a recent discussion (Wigley, Gray, and Kelley 1978). In our opinion such diversity arises for the most part from separation of "early" wood from "late" wood in each tree ring and from separation of lignin from cellulose with its potential for isotope fractionation by wet chemistries. In our work we have usually averaged over three or four rings at a time, and we have always used dry chemistry so that there was no possibility for isotope fractionation.

Thus, in our study of isotope variations in lengthy chronological sequences of tree rings, we are evaluating temperature fluctuations in the sea surface, from which distilled the precipitation which nourished the trees. The sea surface temperatures, in turn, are affected by variations in the ultraviolet spectrum of the sun. The climate record from variations in sea surface temperatures is also stored in the form of variations of organic carbon and uranium in sea sediments. The variations in the organic matter content of the sea core are caused by variations in the amount of living matter growing in the sea surface; this matter becomes more abundant as the sea surface grows warmer but drops to the bottom to form sediments when death occurs. The uranium ions always present in seawater attach themselves readily to organic matter in the seawater and fall to the bottom to form sediments just as does organic matter, so that, when organic matter is more abundant in sediments, uranium is also.

Therefore, we are studying a new technology which may offer the first advance in the evaluation of solar physics since the invention of the solar coronagraph.

Very little is known about fluctuations of the solar constant, even less about fluctuations of the ultraviolet part of the solar constant. Until now, periodicities of the sun have been evaluated solely from the sunspot cycles, observed over 350 years (except for a few desultory observations in ancient China), but the sun has many ways to vary besides sunspots.

As was concluded by the Department of Transportation's Climatic Impact Assessment Program (CIAP), "All that is really known about the solar constant is that it has not changed by as much as a factor of two throughout the ecological history of the earth, that over the past 50 years it has not changed by as much as 10%, and that over the last 10 years, it has not changed by as much as 1%. These known limitations do not rule out long-period (hundreds of years or longer) changes of, say, less than 30% over the entire solar spectrum. Neither do they rule out short-period changes in the near UV nor UV flux (less than 2500 Å) of the order of a factor of two" (Grobecker 1975).

By the method of measuring stable isotope ratio variations in tree ring sequences, we hope to set limits on such solar changes and evaluate their periodicities as far back in time as tree ring sequences exist, for at least 8,000 years, in a sequence tied to the present by overlapping ring patterns onto present patterns. In "floating" sequences in preglacial trees, sequences not fitted to present-day sequences but instead dated by radiocarbon, we have another data base where again evaluation of solar variations becomes possible for the more distant past.

Varves suggest themselves as a rewarding data base from which to evaluate the history of the climate. These are sequences of sediment layers deriving from freshwater streams, the summer sediment having a different particle size and color from the spring sediment, so that the deposit for each year becomes visible as a distinct bicolored layer, and the years of sediment can be counted back from the present. Varve sequences are known which extend over 10,000 years in a single sequence. Their yearly layers contain willow leaves and twigs, as well as wings from beetles, each wing being large enough for a sample for measurement of the hydrogen and oxygen isotope ratios. If the insects have subsisted on rainwater, study of the isotope ratios in their wings should allow one to deduce the history of the climate for the entire varve sequence, as would also measurement of willow leaves.

Sea cores offer a data base which should in principle allow deduction of the history of the local sea surface temperature immediately above the deposition site of the core, for there is enough organic material in sea cores to provide the necessary samples for isotope measurement at frequent intervals versus depth in the core. However, the time resolution is far less accurate than in varves and tree rings, because burrowing sea bottom animals smear the record of the layers.

Any of these data banks, back in the ice ages, can have had their stable isotope ratios perturbed by the huge ice reserves which were removed from the sea and piled up on land, because the ice depletes the oceans in the light isotopes and therefore significantly enriches the sea in the heavy isotopes. Thus sea sediments and continental precipitation, rain and snow, reflect this perturbation as well as perturbations caused by temperature changes alone.

Tree ring sequences extending back from the present to some 8,000 to 10,000 years ago are being prepared by Bernd Becker of the Universität Hohenheim, Stuttgart, and by Dieter Eckstein of the University of Hamburg from huge trees discovered in the beach sediments of the great rivers of Europe, the Rhine, Rhone, and Danube, which grew 10,000 years and more before the present. These scientists fit the tree ring patterns together in overlapping records and have quantities of the authenticated wood dated by pattern recognition to give to analytic laboratories for analysis of climate variations.

We obtained pieces of counted and dated *Sequoia gigantea* from Paul Zinke of the Forestry Department at the University of California at Berkeley, from Henry Michael of the University Museum of the University of Pennsylvania, and from the United States Forest Service in Sequoia–Kings Canyon National Park, California. Sequences of bog oak have been prepared from the ancient oaks dug out of the bogs of England and Ireland by laboratories in those countries.

We hope that tree ring sequences will be prepared from trees of the southern hemisphere, from which one could learn whether climate changes have been simultaneous in both hemispheres. There are large tree stumps in New Zealand, and perhaps similar material could be found in Australia and other southern lands; there may be varve sequences as well in the southern hemisphere.

References

Craig, H., 1961, Science *133*, 1702–1703.

Craig, H., Consiglio Nazionale delle Recherche, Spoleto, Italy, Sept., 1963, 17–53.

Dansgaard, W., 1964, Tellus *16*, 436–467.

Ergin, M., D. Harkness, and A. Walton, 1970, Radiocarbon *12*, 495.

Gould, R. F., ed., 1966, *Lignin Structure and Reactions*, Advances in Chemistry Series 59, American Chem. Soc., Washington, D.C.

Grobecker, A. J., ed. in chief, 1975, *Monograph No. 1*, Climatic Impact Assessment Program, Dept. of Transportation, Washington, D.C.

Herzberg, G., 1945, *Infrared and Raman Spectra of Polyatomic Molecules*, Van Nostrand & Reinhold, New York.

IAEA (International Atomic Energy Agency), 1953–1971, *Environmental Isotope Data, Nos. 1–5*, Vienna.

Keeling, C. D., 1960, Tellus *12*, 200–203.

Olsson, I. U., and M. Klasson, 1970, Radiocarbon *12*, 281–284.
Olsson, I. U., and A. Stenberg, 1967, pp. 69–78 in *Radioactive Dating and Methods of Low Level Counting*, IAEA Symposium, Monaco, Mar. 2–10.
Scpall, O., and S. G. Mason, 1961, Can. J. Chem. *39*, 1934–1943.
Stewart, M. K., and I. Friedman, 1975, J. Geophys. Res. *80*, 3812–3818.
Taube, H., 1956, Ann. Rev. Nucl. Sci. *6*, 277–302.
Vogel, J. C., and J. C. Lerman, 1969, Radiocarbon *11*, 385.
Wigley, T. M. L., B. M. Gray, and P. M. Kelley, 1978, Nature *271*, 92–94.

2. The Experimental Approach

History and Technology

Long-term changes in precipitation, caused by changes in climatic temperature, are well documented in polar ice caps; the heavier of the stable isotopes is depleted in ice laid down in the ice age by comparison with present-day ice. In 1970 we extended this concept to trees, suggesting that they, also, are thermometers. Trees grow from water and atmospheric CO_2. In trees which grow on rainwater, isotope variations in their rings should be climate indicators because the isotope composition in rain and CO_2 varies with temperature.

On May 17, 1971, the Defense Advanced Research Projects Agency funded my proposal that "temperature variations in past climates may be evaluated by measuring stable isotope ratios in natural data banks such as tree rings and varve sequences." I had previously calculated (1972) the theoretical temperature coefficients of the stable isotope fractionations in the manufacture of wood from CO_2 and H_2O, finding that the coefficients are small compared with those measured in rain and snow (IAEA 1969–1971, 1973, 1975).

We considered whether to measure whole wood, lignin, or cellulose, wood being about 25 percent lignin on the average (Gould 1966). But isotope fractionation at climatic temperatures is a function of the frequencies of the chemical bonds (Libby 1972 and references therein). We quote from Gerhard Herzberg as follows: "One would expect the -C-H bond to have essentially the same electronic structure and therefore the same force constant in different molecules, and similarly for other bonds. This is indeed observed" (1945: 192). For the -C-H bonds the vibrational frequencies in lignin and in cellulose are almost equal, but in fact they differ by 6 percent (ibid.: table, p. 195) because cellulose is a multiple alcohol $(H\text{-}C\text{-}O\text{-}H)_n$ and lignin is a polymer containing both aromatic and aliphatic carbons connected to hydrogen. Therefore, assuming thermodynamic equilibrium, variations of the lignin concentration in tree rings might affect the hydrogen isotope ratio by as much as 25 percent of 6 percent, namely, by as much as 1.5 percent.

Likewise, for the -C-O-H bonds, the vibrational frequencies are equal in lignin and in cellulose within a few percent (ibid.). But lignin, different from cellulose, also contains ether linkages, -C-O-C-. The -C-O-H linkage has a C-O bond distance of 1.427 angstroms, and the ether linkage has a C-O bond distance of 1.43 angstroms (Weast 1962). Therefore, the presence of lignin, containing 14 percent oxygen of which 16 percent is ether-linked (Gould 1966; see also fig. 1-6), might affect the isotope ratio of oxygen in whole wood by 0.3 percent × 25 percent × 14 percent × 60 percent, equal to 6×10^{-3} percent, by variation from its average concentration of 25 percent.

The carbon-carbon bond linkage in cellulose is 1.541 angstroms, and in the aromatic groups of lignin it is 1.395 angstroms, a difference of 10 percent (Weast 1962). Therefore, the presence of lignin, containing 75 percent carbon (Gould 1966) of which 60 percent is aromatic, may affect the carbon isotope ratio by 10 percent × 75 percent × 60 percent × 25 percent, or about 1 percent, by variation from its average concentration of 25 percent.

On these numerical arguments and on the necessity to avoid isotope exchange with liquids, we based our decision to measure stable isotope ratios in whole wood.

The next problem concerned which trees to measure. Many tree ring sequences can be counted with an accuracy of about 1 year. Those which are not yet tied to modern sequences by overlapping ring patterns (said to be "floating") can be dated in favorable cases with an accuracy of about 30 years by radiocarbon, depending on the age and the number of radiocarbon measurements which are made.

But, to prove our hypothesis that trees are thermometers, we needed to compare our measurements of stable isotope ratios in the tree rings with mercury thermometer records near where the trees grew. Thus we could not use bristlecone pines, because there is no lengthy temperature record for hundreds of miles near their home in the White Mountains of California (nevertheless, some measurements of D/H in bristlecone pines have been published; see Epstein and Yapp 1976).

Because the longest temperature records are in Europe, we obtained a German oak from the only tree ring laboratory then existing in Europe, the laboratory of Bruno Huber in Munich (Huber and

Siebenlist 1969), where the oak rings were counted and labeled with the number of the years in which they grew. More recently his students, Bernd Becker and Dieter Eckstein, who have established tree ring laboratories in Stuttgart and Hamburg respectively, have sent us additional sequences of German oaks in which they have counted and labeled the rings.

From K. Y. Kigoshi in Tokyo, we obtained a 2,000-year ring sequence of a cedar from the southern tip of Japan, in which Kigoshi had counted and labeled the rings; in addition, he verified his dates by making fifty radiocarbon measurements in its wood. Although radiocarbon dating's accuracy is only about 40 years, by making fifty measurements the verification achieves an accuracy of $40/(50)^{1/2}$ or about 6 years.

The temperature records needed for comparison with the German oaks exist at nearby Basel and Geneva, extending back more than 2 centuries, and in central England, extending back 3 centuries (Manley 1953, 1959, 1974). Temperature records for the cedar have existed at Miyazaki, Japan, since 1890. In addition, there are surrogate climate records for the Far East in the form of records of dates of cherry tree bloomings, of number of days per year when lakes were frozen, and of number of snowy days per year (Libby et al. 1976).

The reason for our use of local air temperature records is that sea surface temperature records are not readily obtainable except in the commercial sea-lanes. But we know that oceans comprise 80 percent of the surface of the world and determine the air temperatures of the continents. From the evidence of tritium in rainwater, we know that rain makes three or four "hops" raining out and evaporating in crossing a continent, so that rainwater retains the signature of the isotope content caused by the temperature of evaporation from the sea surface, and this temperature is intimately tied to local air temperatures where the tree grew.

We had an idea of the magnitude of the oxygen and hydrogen isotope variations we could expect to find in these trees because, since 1953, the International Atomic Energy Agency in Vienna (IAEA) has measured and published the stable isotope variations in rain and snow, month by month, versus air temperature, for 155 worldwide weather stations, including those in Germany, Austria, and Japan (see fig. 1-1).

Carbon isotope fractionation by CO_2 absorption at the air-water interface has been measured (Degans et al. 1968), and the ratio $^{13}C/^{12}C$ has been found to vary from 9.2 to 6.7 ppt over the temperature range 0° to 30° C. Thus we may write the enrichment E at time t as follows (derived in Libby 1973):

$$E_t = 0.027\,(M/M^*)\,(1 - 0.5\times10^{-3}\,(T - T^*))\,(1 + 8.5\times10^{-3}\,(T - T^*)) \quad (1)$$

where the number 8.5×10^{-3} is taken for the average fractionation of carbon-13 at the sea-air interface and where T is the average temperature in degrees C at time t, M is the mass of the biosphere at time t, T^* is the average temperature today (assumed to be 20° C), and M^* is the mass today.

The values of E_t so computed are listed in table 2-2. The correction for fractionation of carbon dioxide at the sea surface is a serious one. It makes the interpretation of $^{13}C/^{12}C$ variations in wood difficult and militates against the use of the isotope ratio of carbon as a thermometer. This correction, when applied to variations of carbon-14 in wood, is able to explain the Suess radiocarbon "wiggles" of about 100 years' duration each, without the need to invoke changes in the neutron flux from the sun (Libby 1973).

We considered whether old heartwood could exchange isotopes with modern sapwood. On the contrary, there is compelling evidence that when sapwood passes into heartwood it becomes sealed against sap and therefore against isotope exchange with sap, at least in species having tight rings. For example, Huber has shown, using biological dyes in many tree species, that water conduction remains limited to the outermost annual ring.

There is additional evidence that radiocarbon sugars, injected in or ingested by a living tree, do not move into neighboring rings but remain in the ring in which they were injected or ingested. Furthermore, we have observed that, when an intact block of California redwood was soaked for a month in an atmosphere saturated with water previously labeled with delta D_{SMOW} = 1,170 ppt, no deuterium exchange with wood was observed. This is reasonable to expect because wood is a remarkable polymer, containing very large molecules, cross-linked internally and to each other with hydrogen bonds.

Hence we concluded that the climate record in heartwood cannot be modified or perturbed by the sap in the outermost ring of the current active year.

For our first tree sequence (Libby 1974; Libby and Pandolfi 1973a and b, 1974, 1976), we measured D/H by reacting sawdust with uranium to produce H_2, 99 percent quantitatively. For measurement of $^{18}O/^{16}O$, we modified the method of D. Rittenberg and L. Pontecorvo (1956) by carrying it out at very high temperatures, 99 percent quantitatively. The temperature must be 525° C. If it is lower, the reaction is not quantitative; see the section on our chemistry later in this chapter. To measure the stable isotope ratio in carbon, we burned sawdust to completion in oxygen.

Whether the oxygen in tree rings comes from water or from CO_2 is not in question, because M. Cohn and H. Urey (1938) showed that isotopic equilibrium between the two substances is obtained in a damp atmosphere within a few hours at room temperature.

For measurement of the oaks we used, perforce, a mass spectrometer of somewhat low accuracy, and we achieved the accuracy to demonstrate that trees are thermometers by making many measurements on each sample. On the tree sequences which we measured later, we used high-precision spectrometers with accuracies of ±0.1 ppt for $^{18}O/^{16}O$ and $^{13}C/^{12}C$ and ±2 ppt for D/H. The measurements are expressed in terms of δ_D and δ_{18}. In an intermediate stage, we used the original mass spectrograph built after World War II by Harold Urey at the University of Chicago, later relocated at the University of California at San Diego, and we were grateful that Harold Urey and Kurt Marti allowed us access to it.

Sample Preparation

We mill a groove perpendicular to the tree rings, that is, along the radius of the tree; sawdust from each few rings is collected into an individual vial with a camel's hair brush. The vials are dried at 50° C and capped off to protect the dried sawdust from damp air. Our chemistry is now as follows.

To evolve CO_2 for measurement of $^{18}O/^{16}O$: pump for 4 hours on 3 milligrams of sawdust mixed with 120 milligrams of $HgCl_2$ *in vacuo*. Seal the container. Heat at 525° C for 4 hours; if the temperature is less than 525° C, production of CO_2 does not quantitatively remove oxygen. React with triple-distilled quinoline at boiling temperature until the quinoline turns yellow. Freeze in a slurry of ethanol and dry ice at −120° C. Pass the gas through two traps of dry ice and acetone.

To evolve H_2 for measurement of D/H: burn 5 milligrams of dry sawdust in 1 atmosphere of O_2 in a cupric oxide furnace at 750° C. Use oxygen purified over silica gel and cupric oxide to insure that the O_2 is hydrogen-free. Freeze out H_2O and CO_2 in a liquid oxygen trap. Release CO_2 at dry ice temperature. React H_2O vapor on clean uranium shavings at 950° C, thus producing H_2 quantitatively.

To evolve CO_2 for measurement of $^{13}C/^{12}C$: burn 3 milligrams of sawdust in dry, clean oxygen gas.

We thank Willard Libby for advising us how to make this dry chemistry quantitative.

Recent Trees and Thermometer Records

To prove that trees are thermometers, we needed to measure modern trees whose rings have been correctly counted and to obtain lengthy mercury thermometer records from places near where the trees grew, perferably from the same altitude. The oldest European temperature records in the *World Weather Records* are for Basel and Geneva, each at about a 300-meter altitude. We fortunately obtained from Huber's laboratory a German oak grown in central Germany at an altitude of about 300 meters (fig. 2-1), which had been correlated with the fiducial oak tree ring sequential pattern developed by Huber and his colleagues. Later, we obtained samples of similarly calibrated German oaks from Becker and Eckstein.

So far, we have measured isotope ratios in four modern trees of four different species at four different altitudes, latitudes, and longitudes and compared them with local temperature records from mercury thermometers: German oak, *Quercus petraea*; Bavarian fir, *Abies alba*; Japanese cedar, *Cryptomeria japonica*; and California redwood, *Sequoia gigantea*. For the parts of the oak and the cedar extending beyond the beginning of mercury thermometer records, we compared the measured isotope ratios with surrogate evidence of climate change, such as lateness of cherry tree bloomings, number of snowy days per year, and number of days per year when lakes were frozen. We found significant correlations.

In figure 2-2 we compare carbon, deuterium, and oxygen isotope measurements for a German oak with English temperature records (Manley 1953, 1959, 1974) back to the time of the invention of the mercury thermometer in the seventeenth century. Temperature records of Basel and Geneva from 1750 resemble those of central England and thus fit the isotope ratios well, but we show the English thermometer records here because they extend further into the past. The carbon isotope ratios for the years 1890 to 1950 have been corrected for the effect of fossil CO_2 production (Suess effect); the maximum correction, that for the year 1950, was taken as 8.4 percent for

FIG. 2-1. The German oak slice from the Munich laboratory of Bruno Huber, showing the milled groove.

a maximum increase of 2.1 ppt in $^{13}C/^{12}C$ because, in wood from rings of 1920, two radiocarbon dates were measured by Rainer Berger, University of California at Los Angeles, as 375 ± 35 years old with respect to 1950, whereas the actual age is 1950 − 1920 = 30 years, corresponding to 4.2 percent dilution of atmospheric $^{14}CO_2$ by inert CO_2 produced by the burning of coal and oil up to 1920. In 1950 the correction for ^{14}C in this particular tree should be 8.4 percent, and the correction for ^{13}C dilution should be 9.4 percent of 25 ppt or 2.35 ppt.

Direct measurements of $^{18}O/^{16}O$ in rain and snow are available in IAEA publications (1969–1975); figure 2-3 shows the δ_D and δ_{18} correlations with temperature for Austrian stations for 15 years. The

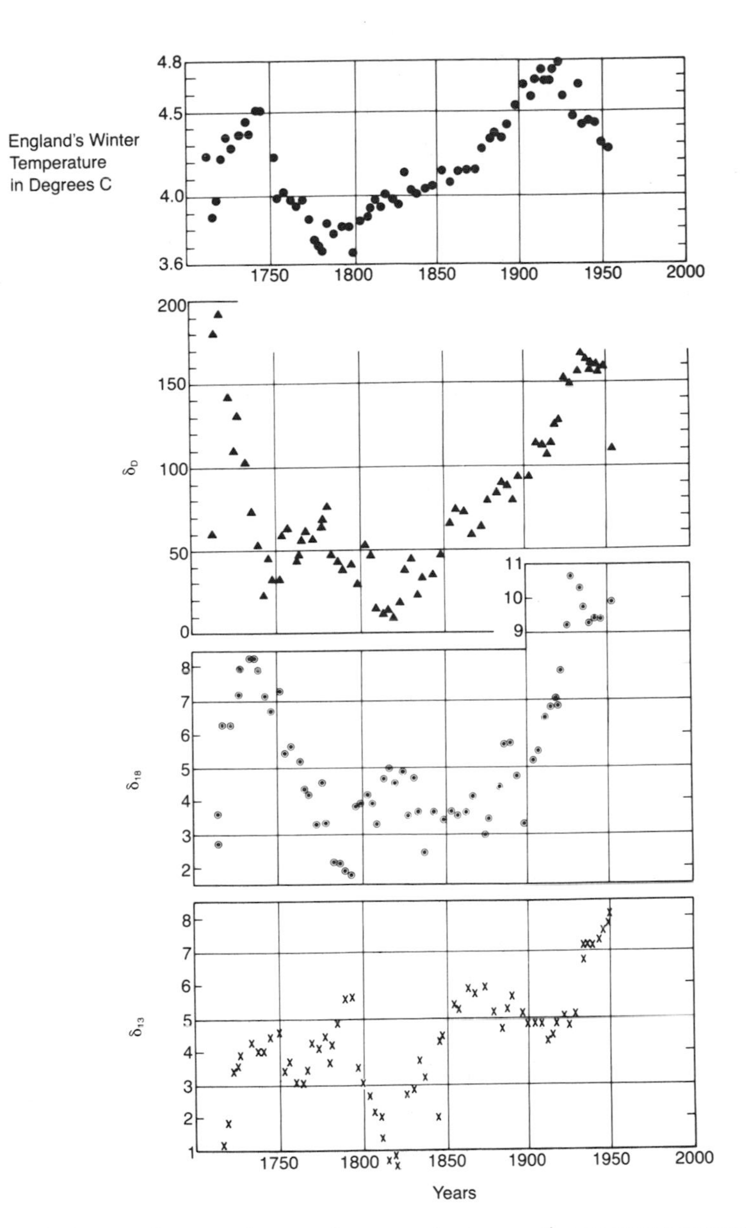

FIG. 2-2. Deuterium, carbon, and oxygen isotope ratios versus temperature in a German oak, 1650 to 1955.

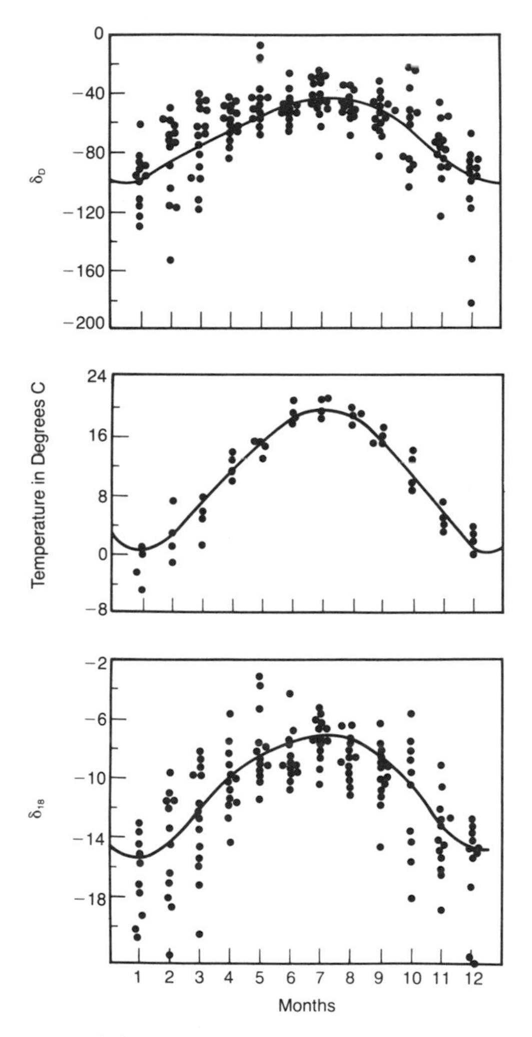

FIG. 2-3. Oxygen and deuterium isotope ratios versus temperature in Austrian precipitation, 1960–1975. The drawn curves are phenomenological.

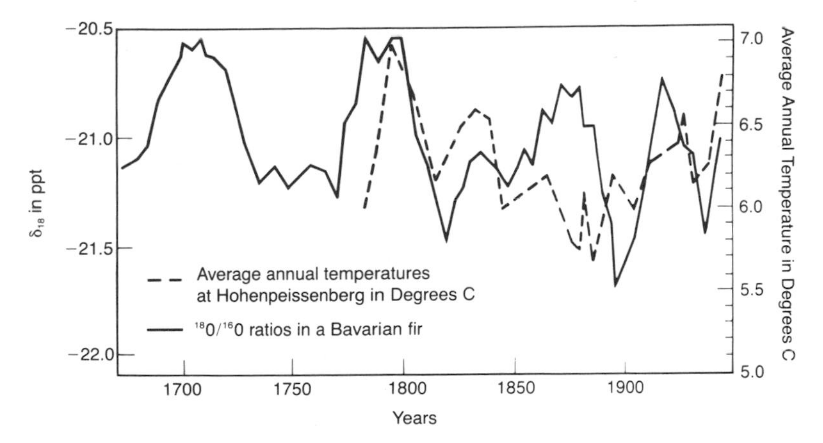

FIG. 2-4. Oxygen isotope ratios versus temperature in a Bavarian fir. The lines are phenomenological.

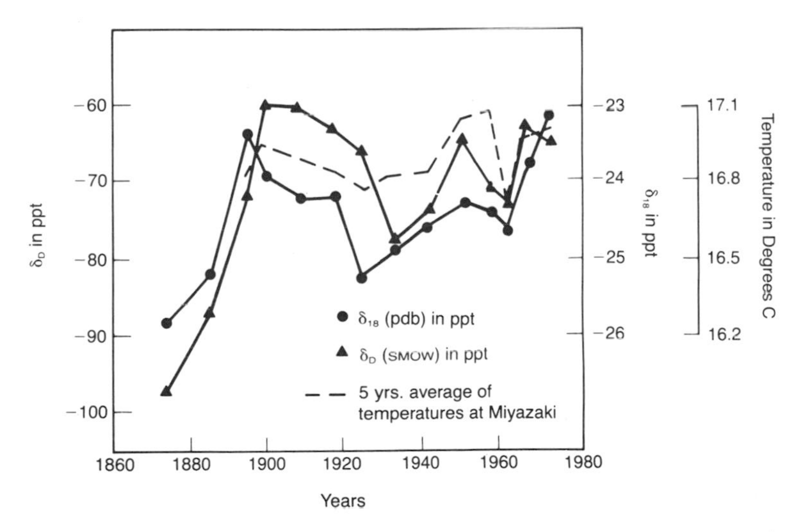

FIG. 2-5. Oxygen and deuterium ratios versus temperature in a modern *Cryptomeria japonica*. The lines are phenomenological. The notation pdb refers to a standard made of a ground-up fossil belemnite snail from the Pee Dee formation, a geologic sediment; the notation SMOW refers to standard mean ocean water.

isotope concentration of precipitation varies similarly with temperature in many other places, as shown by plots of the IAEA isotope measurements against air temperature.

Figure 2-4 shows the $^{18}O/^{16}O$ ratio for a Bavarian fir, *Abies alba*, the rings of which were counted by Becker and Siebenlist (1970), compared with temperature records made near where the tree grew, both coming from 1,000 meters on the north slopes of the Alps. Since this is mountainous country, local differences in temperature may be expected.

Figure 2-5 shows δ_{18} and δ_D in a *Cryptomeria japonica* from Yaku Island compared with a local temperature record, in southern Japan, from the weather station at Miyazaki. The tree grew at 1,350 meters, whereas Miyazaki is at sea level.

Figure 2-6 shows δ_{18} in a modern part of a *Sequoia gigantea* which grew in the giant forest of Sequoia–Kings Canyon National Park, Three Rivers, California, at 1,940 meters (Daugherty 1976; Michael 1976), compared with air temperatures for the same years measured in Yosemite National Park, about 100 miles south, at an altitude of 1,200 meters. We compare with summer air temperatures

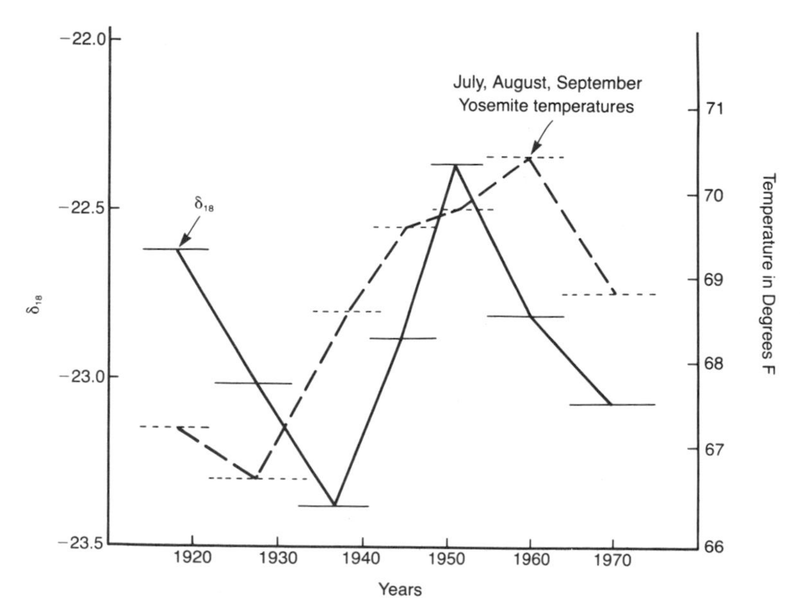

FIG. 2-6. Oxygen isotope ratio versus temperature in a modern *Sequoia gigantea*. The lines are phenomenological.

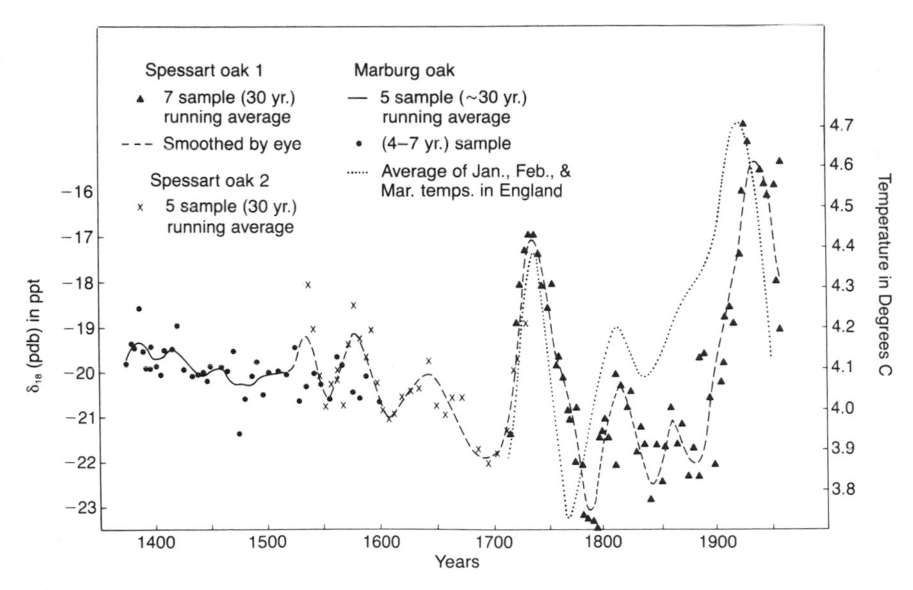

FIG. 2-7. Oxygen isotope ratios in a sequence of German oaks, 1350 to 1950. The drawn curves are phenomenological.

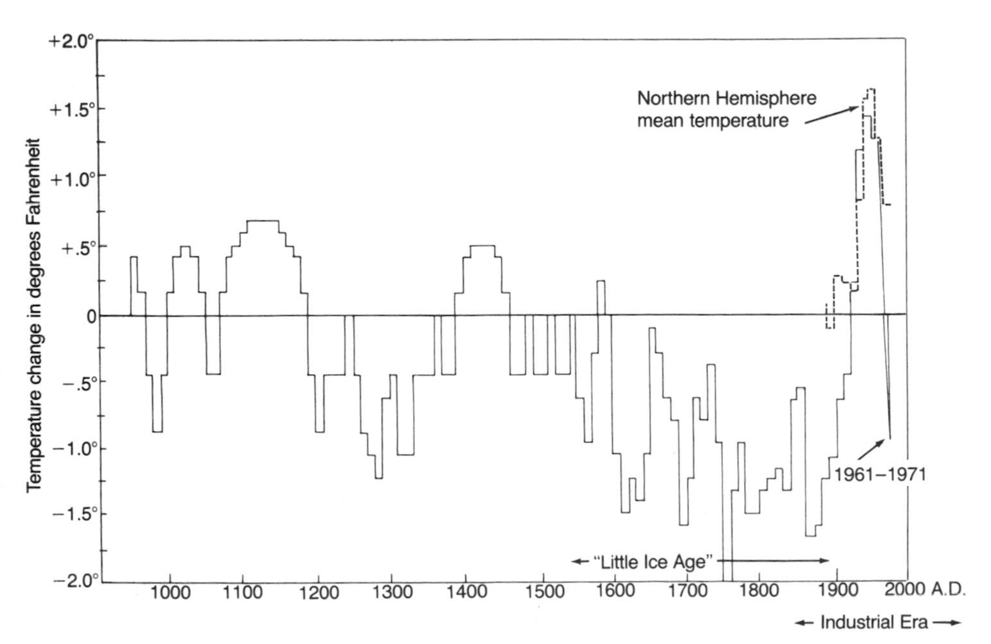

FIG. 2-8. A thousand-year history of Iceland's temperature.

because the sequoias in Kings Canyon grow for the most part in summer, when many inches of rain fall from clouds coming from the Gulf of Mexico.

Old Trees and Surrogate Evidence

In figure 2-7 we show oxygen measurements for German oaks which grew before thermometers were invented, indicating warm intervals in 1530, 1580, and 1650 and cold periods at about 1700 and 1800 in agreement with P. Bergthorsson's deductions of climatic variations in Iceland (1962) (fig. 2-8) and in agreement with the historical evidence of severe climate deteriorations in the First and Second Little Ice Ages in Europe.

In figures 2-9 and 2-10 we show oxygen and deuterium isotope ratios in a *Cryptomeria japonica*, also from the island of Yaku at the southern tip of Japan as in the modern cedar (fig. 2-5), for the years

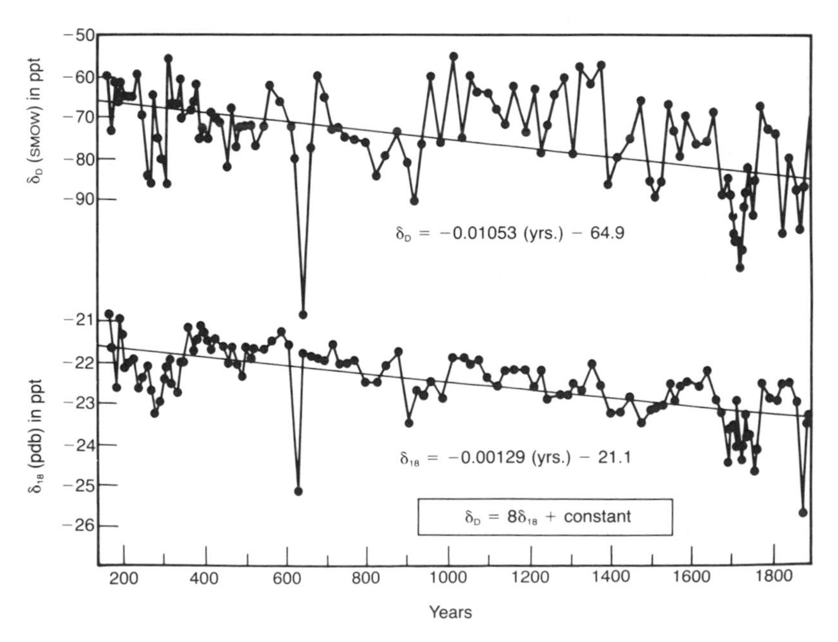

FIG. 2-9. Deuterium and oxygen isotope ratios in a *Cryptomeria japonica* from Yaku Island, Japan, 160 to 1900. The drawn fits are phenomenological. The slopes are determined by computerized least squares fits to the measured points.

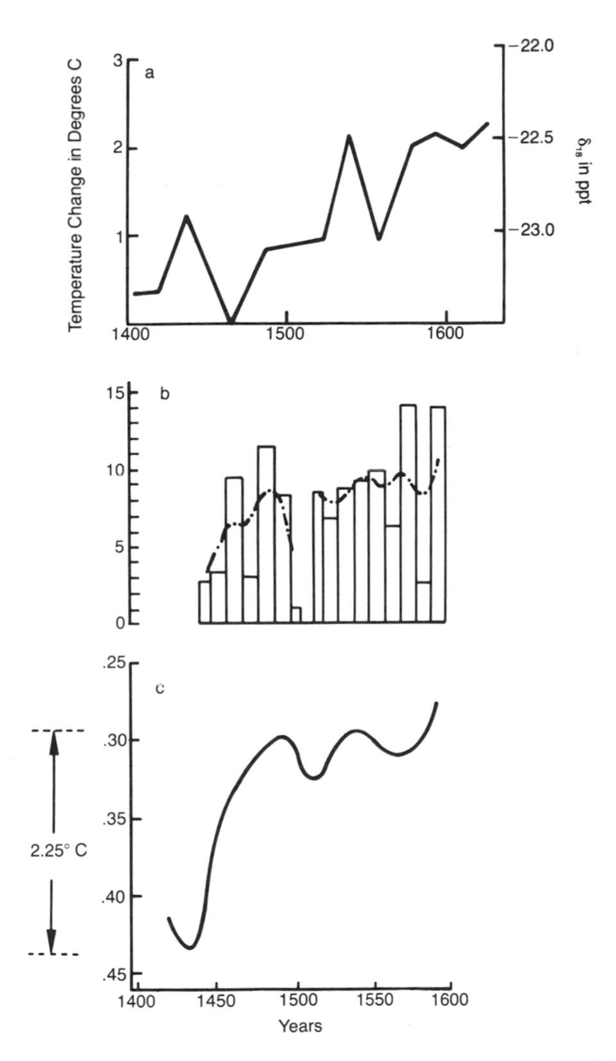

FIG. 2-10. Oxygen isotope ratios in *Cryptomeria japonica* calibrated for temperature change, obtained from (a) time of cherry tree bloomings, (b) first freezing of Lake Biwa, and (c) number of snowy days per year. The drawn curves and lines are phenomenological (Yamamoto 1971; Chu 1973).

160 to 1900, compared with temperatures deduced from old diaries for Japan and China.

In figure 2-11 we show the oxygen isotope ratios for the same *Sequoia gigantea* (fig. 2-6) for the years 1750 to 1975. This shows the same negative slope indicating deteriorating climate as does the Japanese cedar in figures 2-9 and 2-10, amounting to a long-term decline of about 1.6° C in Miyazaki in the last 2,000 years. This is significant, especially considering that there was a decline of about 10° C in the last ice age.

Climate Periods

We have made Fourier transforms (Blackman and Tukey 1958) of the data in figures 2-9 to 2-11 to deduce the power spectra of periodicities. The results are shown in figures 2-12, 2-13, and 2-15. The same periods are found in deuterium and oxygen in the Japanese cedar, within experimental error, as indicated from the widths of the peaks of the power spectra in figures 2-12 and 2-13; they are listed in table 2-1.

Our samples consist of wood from about 5 years each, so we cannot expect to find meaningful evidence for periods of less than about 40 years in these data. The Japanese cedar spans only about 1,800 years, so we cannot ask for meaningful evidence for periods of more than about 250 years. In making the Fourier transforms, we have used the deviations of the isotope ratios from the long-term slopes—that is, we have corrected for the slopes. We have tested the meaningfulness of the periods in table 2-1 by manufacturing data sets consisting of a number taken from the least squares fit to the data, namely, the straight-line fit, plus a random number varying over the numerical range of the deviations from the line. Each set of manufactured data was subjected to Fourier transform. In each case, in the transforms for thirty such manufactured data sets, no peaks were generated of significance, indicating a confidence level for the periods in table 2-1 of better than 96 percent.

The Fourier transforms were performed in the standard way. No smoothing or filtering was employed. Subtraction of the data from the least squares fit removes the constant or linear term characterizing a Markovian process. Fourier transform of the differences from the linear fit suppresses the enhancement of both the power and the amplitude spectra at low frequencies.

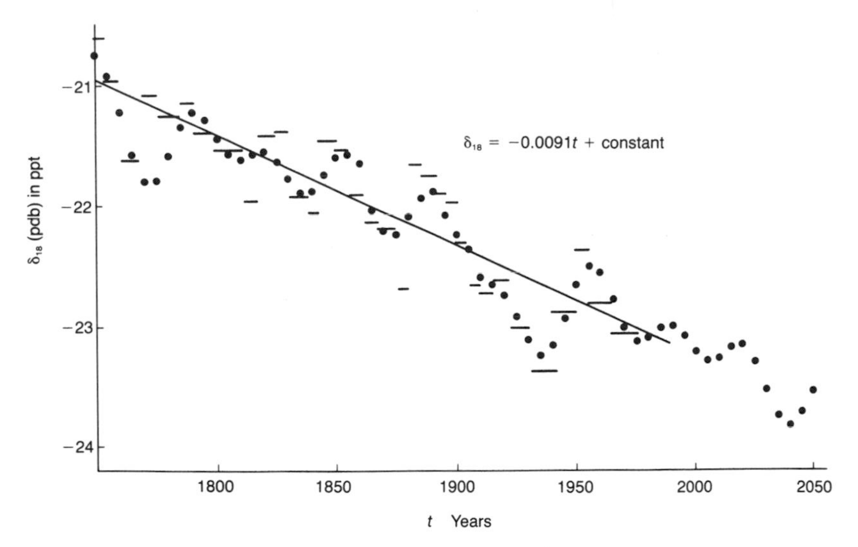

FIG. 2-11. Oxygen isotope ratios in 225 years of a *Sequoia gigantea* ring sequence from California. The horizontal lines represent measured values. The black circles represent values predicted from the reconstruction using evaluated periods, phases, and amplitudes obtained from Fourier transform. The prediction has been projected into the future.

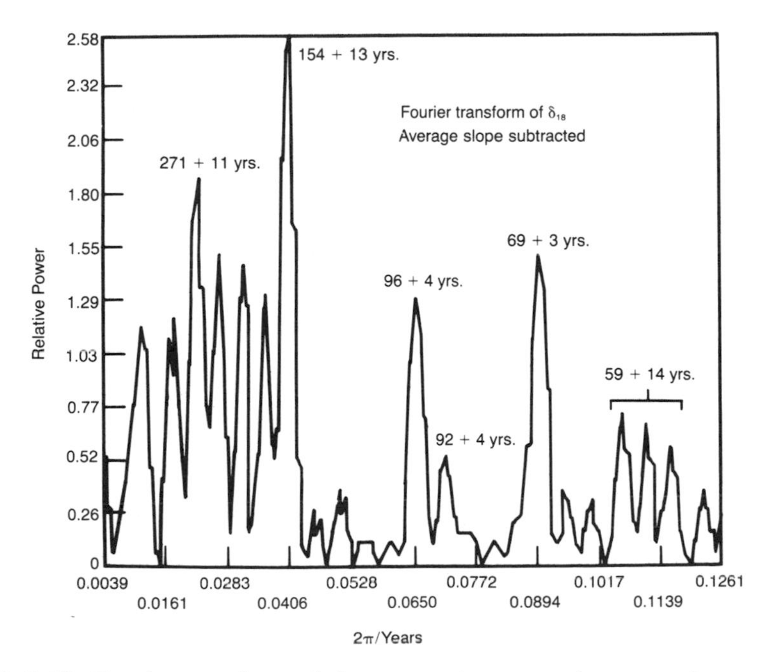

FIG. 2-12. Fourier transform of the oxygen isotope ratio versus time, measured in *Cryptomeria japonica,* into power versus reciprocal period.

The periods derived from the transform of the oxygen isotope ratio in the *Sequoia gigantea* appear to agree with those in the *Cryptomeria japonica*, as shown in table 2-1. In the sequoia, the samples were taken for as few as 2 years at a time, so that it is possible to have confidence that the 33-year period found in it is meaningful. But, because the tree has been analyzed for only about a 225-year span, we cannot ask for meaningful evidence of periods of more than 143 years; and, although it appears to agree with the period of 157 years found in the *Cryptomeria japonica*, and although in our experiments with artificially generated isotope ratios in the Fourier transforms no such large amplitudes were produced, still one must take caution.

In the Fourier transforms, amplitudes and phases are generated for each period. We have put the periods, amplitudes, and phases back into the summation of transcendental functions which represents the function of the data, and we have used this expression to generate the original data (see table 2-1), as shown by the black circles in figure 2-11. By running this function into the future, we have made a prediction of the climate to be expected in Kings Canyon: the climate will continue to deteriorate on the average. Superim-

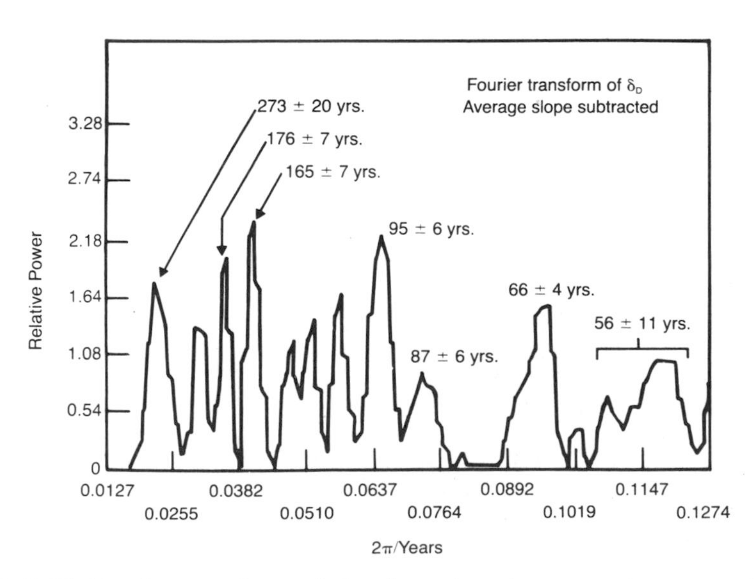

FIG. 2-13. Fourier transform of the deuterium isotope ratio versus time, measured in *Cryptomeria japonica*, into power versus reciprocal period.

Table 2-1. Periods Yielded from Several Sets of Measurements by Fourier Transform (Years)

Cryptomeria japonica		Santa Barbara Sea Core		Bristlecone Pine	Greenland Ice Well
D/H	$^{18}O/^{16}O$	organic carbon	uranium	carbon-14	$^{18}O/^{16}O$
156	156	161	156	162	179
110	124	121	118	108	—
97	97	95	95	—	100
86	88	82	81	—	78
65	70	71	70	—	68
58	55	55	53	—	55

posed on the long-term decay there will be a temporary warming-up followed by a greater rate of cooling-off.

Naturally, a complete analysis of the 3,000-year span of the *Sequoia gigantea* which we have in hand will yield a more reliable prediction of the future climate for Kings Canyon.

Five of the periods found in the oxygen and hydrogen isotope ratios of the Japanese cedar are also evidenced in variations of the oxygen isotope ratio versus depth in the Greenland ice (Dansgaard et al. 1971), as listed in table 2-1 (Libby and Pandolfi 1977). The remarkable agreement between our tree records of oxygen and hydrogen isotopes and the ice record of oxygen isotopes shows itself in yet another way: we have found the D/H and $^{18}O/^{16}O$ ratios for the Japanese cedar to be significantly correlated in opposite phase to the ^{14}C variations in bristlecone pines of southern California (Libby et al. 1976), as measured by Hans Suess (1973). Similarly, W. Dansgaard and his colleagues (1971) found the oxygen isotope record in Greenland ice to be significantly correlated in opposite phase with bristlecone carbon-14. Furthermore, the Fourier transform of the carbon-14 variations shows four of the same periods as in our tree and in the Greenland ice; see table 2-1.

We conclude from the correlations of these four sets of data that the calculation of Dansgaard et al. of the age of ice versus depth in the Greenland ice cap seems to be correct with an error of not more than a couple of years, at least over the last 800 years. We conclude that the climate variations in Greenland, southern Japan, and southern California have had the same periodicities for the last 800 years or more.

A logical explanation for the global nature of these correlations is that they are all related to variations of the sun, which cause variations in the temperature of the sea surface, thus causing variations in the isotopic composition of water vapor which distills off the sea and is stored as wood in trees and also forms the annual layers of the ice cap. The variations of the sun are furthermore related to the flux of solar neutrons in the earth's atmosphere and so caused small variations in the carbon-14 content of the bristlecones. During times of a quiet sun, the average carbon-14 production is about 25 percent greater than when solar activity is high (Suess 1973).

We now report two additional data sequences (Kalil and Kaplan 1976) versus age, in which the same periodicities are revealed, namely, the organic carbon and uranium concentrations in a sea core from the Santa Barbara Channel off California. Preservation of annual varves in this anoxic environment provides a record of the age of the sediments, there being one varve or distinct layer for each year. The concentrations, versus depth, of organic carbon and uranium were measured in a continuous sequence of samples, each containing 7 years of sediment, in sea core PT-8G, spanning the years 1264 to 1970. The age of sediment versus depth in the core was determined by comparison of its varves with those in core 214, in which the varves had been counted (Doose 1978; Soutar and Isaacs 1969).

The age determination allowed Fourier transforms to be made, transforming concentration versus depth into signal power and amplitude versus the period expressed in years. The periods found are listed in table 2-1. The sea core spans over 700 years, with each sample containing 7 years of sediment, so that evidence for periods between 40 years and 200 years should be meaningful (Pandolfi et al. 1978).

Table 2-1 shows that six periods found in uranium and organic carbon variations in the sea core are also found in stable isotopes of hydrogen and oxygen in the tree and in oxygen isotope variations in the ice, enriching the interpretation of climate variations on a global scale and enriching the attribution of these periodicities to variations of the sun causing changes in sea surface temperature. The temperature of the sea surface determines the abundance of life and therefore the abundance of organic matter which falls to the ocean bottom and binds uranium ions in the seawater to it as it falls. (It is well known that in uranium ores the amount of organic carbon is proportional to the amount of uranium.)

The recently completed Climatic Impact Assessment Program

(CIAP) study of the stratosphere (Grobecker 1975) indicated mechanisms by which the sun's variation may influence temperature on the earth's surface. In particular, solar energy absorbed in the stratosphere is rapidly converted by chemical reactions, producing a variety of chemical species which reradiate both to space and to the ground. Thus climate on the ground is sensitive to variations in stratospheric chemical components. The solar light is known to vary by factors of 2 and 3 in intensity at short wavelengths, affecting the concentrations of those species. Reradiation to the ground is especially sensitive to ozone and water concentrations in the stratosphere, and those variations are only just beginning to be assessed. Study of stable isotopes in trees and of uranium and organic carbon in sea sediments as a function of time in the past may be a powerful way of examining past misbehaviors of the sun.

The most well known misbehavior, that is, variation, of the sun is seen in the number of its sunspots, with a change in number between the north and south hemispheres of the sun every 10.5 years and a completed magnetic sunspot cycle every 21 years on the average. Supposing that the sun's variations have caused the stable isotopes in our tree, the stable oxygen isotopes in the ice cap, the organic carbon and uranium in the sea core, and the carbon-14 in the bristlecone pines all to vary with the same periodicities, then it seemed reasonable to us to look for evidence of the 10.5- and 21-year sunspot cycles in our trees. In order to find meaningful evidence for such short periods, it is necessary to measure the stable isotope ratios in every single year of a tree ring sequence.

Accordingly, our student, Paul Hurt (1978), measured the oxygen isotope ratios in a ring sequence of 72 years in a B.C. sample of *Sequoia gigantea*. He chose this sequence because the rings are unusually wide and consequently easy to separate. He analyzed it year by year; his results are shown in figure 2-14. The Fourier transform of the ratios, shown in figure 2-15, has significant power signals at 6.69, 10.4, and 37.4 years. To determine the significance of these signals, we generated thirty random number sequences of seventy-two numbers each having values between −18 and −21 ppt, the range of the values of the measurements in figure 2-14, and made their Fourier transforms, but none of these computer experiments produced peaks with more than half the amplitude of those in figure 2-15. This indicates that the periods of 6.69 and 10.4 years are real, with a confidence of 96 percent or better (no peaks in thirty experiments with random number sets).

The reality of the large peak at 37.4 years might be questioned

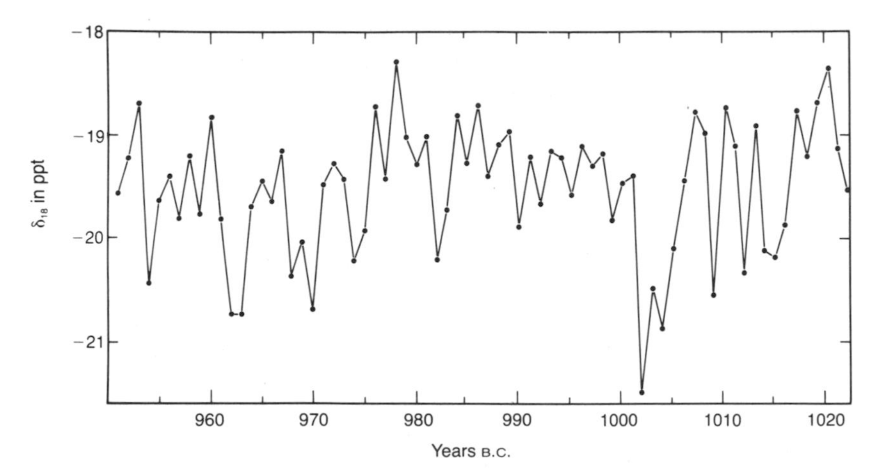

FIG. 2-14. Oxygen isotope ratio versus time in *Sequoia gigantea*, measured year by year for a 72-year period.

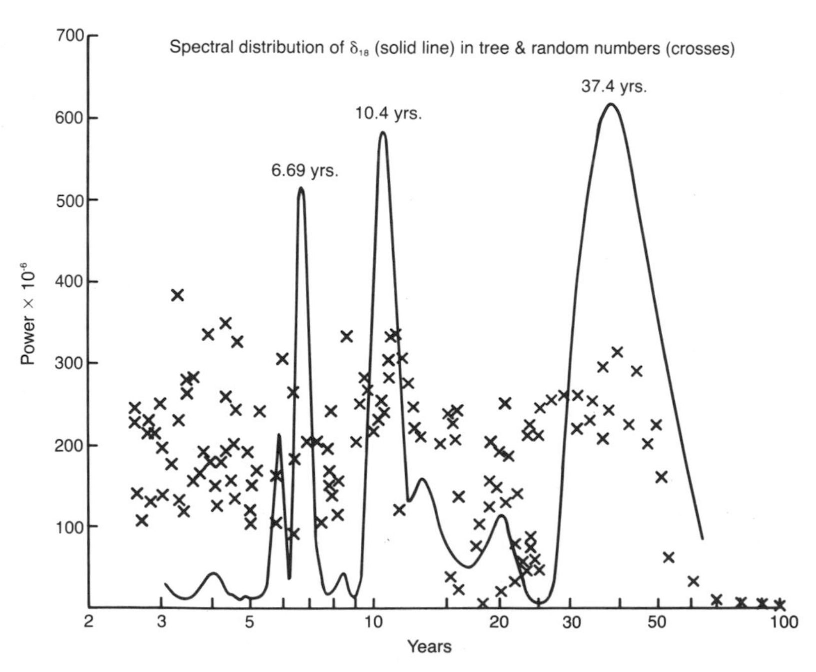

FIG. 2-15. Fourier transform of the oxygen isotope ratio, measured year by year in *Sequoia gigantea*, into power versus period.

because 37.4 years is a large fraction of 72 years. However, it appeared in the Fourier transform of the oxygen isotope ratios of the modern part of the same sequoia (1750–1975) as 33.4 years, so we believe it to be real (see figs. 2-11 and 2-16). Furthermore, we cut the data sequence from 72 years to 63 years, made its Fourier transform, and found that the 37-year signal remained unaffected, indicating that it is not an artifact of a half harmonic of 72 years.

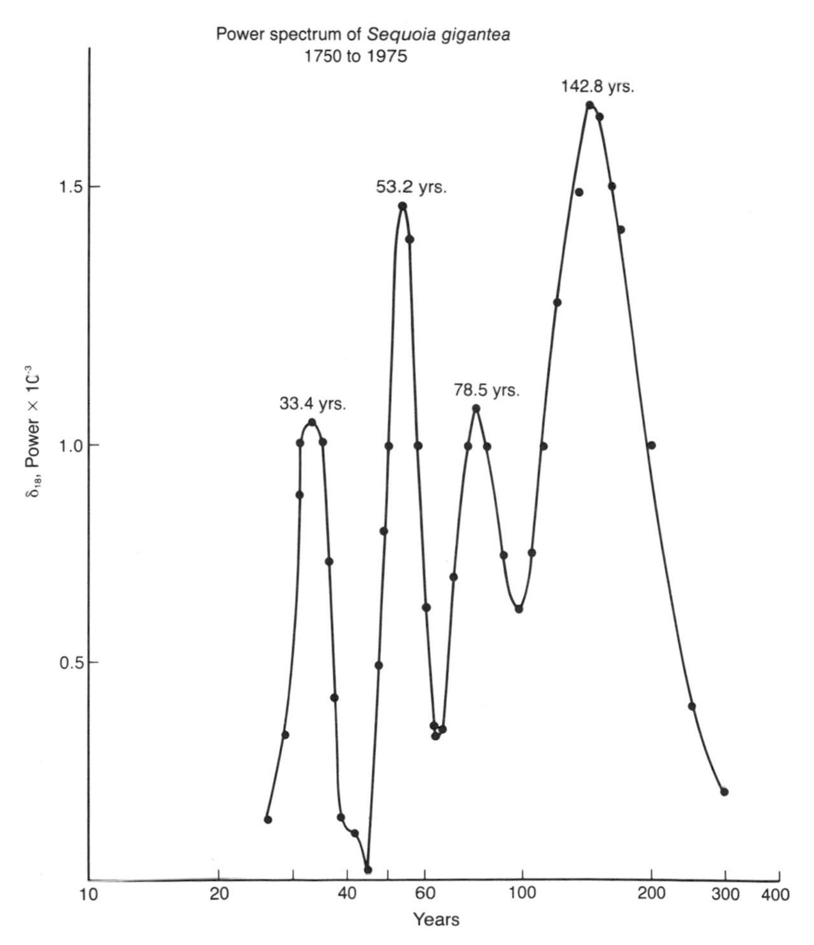

FIG. 2-16. Fourier transform of the oxygen isotope ratio in 225 years of a modern *Sequoia gigantea*, measured 5 years at a time, into power versus reciprocal period.

For comparison with the periodicities in the sunspot cycle, we have made Fourier transforms of the over 350 years of the sunspot counts (Waldmeier 1961). By assuming all sunspot counts to be positive numbers (Cohen and Lintz 1974), we obtained the power spectrum shown in figure 2-17 with signals at 9.94, 10.47, and 11.06 years, for a strong signal averaging to 10.6 years. By assuming sunspot counts in the northern hemisphere of the sun to have a magnetic polarity opposite to that in the southern hemisphere (Hill 1977), we obtained the power spectrum shown in figure 2-18 with a strong signal at the well-known period of 21 years.

The fact that the oxygen isotope power spectrum in figure 2-15 does not show a significant signal at 21 years but shows a strong signal at 10.4 years (which could easily be 10.5 years within our errors) we interpret as indicating that the influence of the sunspot cycle on the earth's climate is effected by neutral particles (which are affected by the number of spots but not by their magnetic polarity, because

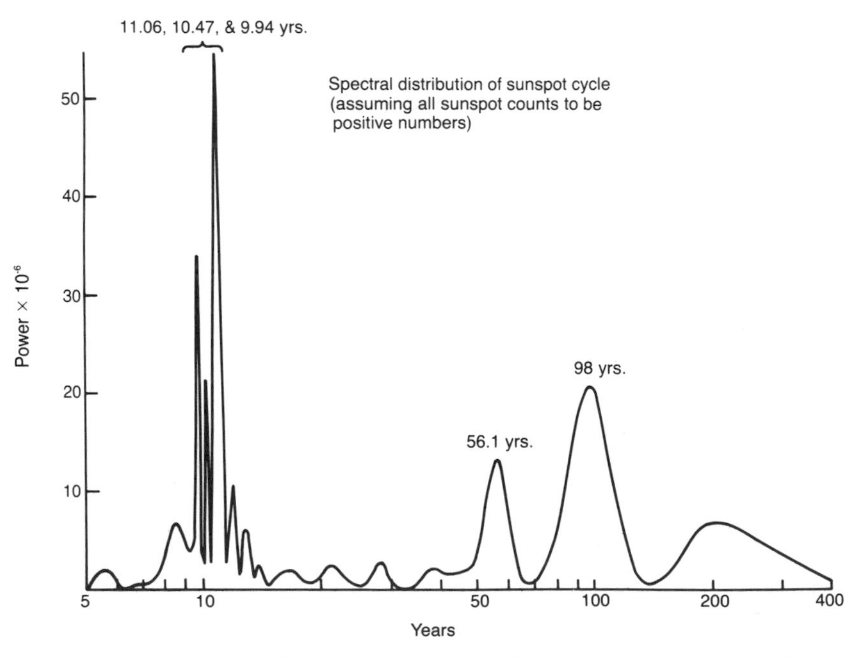

FIG. 2-17. Fourier transform of sunspot numbers versus time, counted year by year for more than 350 years, into power versus period, assuming all sunspots to be identical in magnetic polarity.

they are neutral), probably photons. Neutral particles are not deflected by the earth's magnetic field, so they are able to come to the earth symmetrically from either hemisphere of the sun.

The signals in the oxygen isotope record from the sequoia's rings at 6.69 and at 37.4 years may or may not be related to periodicities of the sun. The sun has many ways to vary, apart from the sunspot cycle, such as fluctuations in frequency of solar flares and plages and misbehaviors of the overall magnetic field of the sun and of the solar corona. And there are also geometric effects.

The conclusions of Hurt's study of year-by-year oxygen isotope ratios in 72 years of *Sequoia gigantea* are thus supportive of the conclusions of the CIAP study (Grobecker 1975) that solar variations influence the abundance of many kinds of chemical species in the stratosphere, and therefore influence the amount of solar energy they absorb and reradiate to earth, and therefore influence the surface temperature of the earth and especially the surface temperature

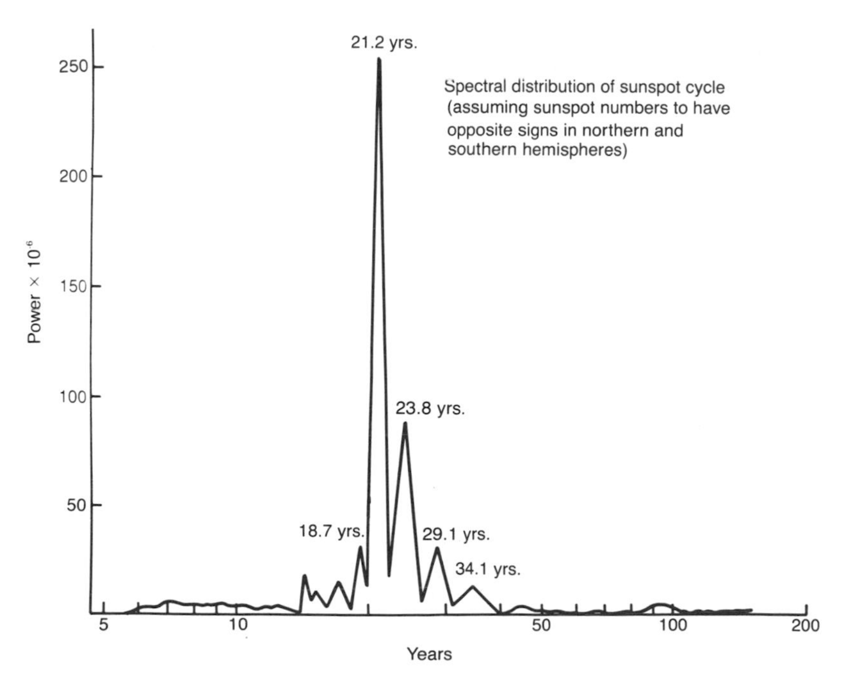

FIG. 2-18. Fourier transform of sunspot numbers, counted year by year for 200 years, assuming sunspots to have opposite magnetic polarity in the northern and southern hemispheres of the sun, into power versus period.

of the oceans. It is the surface temperature of the oceans which produces the phenomena we have discussed: the isotope ratio variations in rain and hence in tree rings, in the Greenland ice cap, and in the organic carbon and uranium concentrations in sea cores. Furthermore, variations in the sea surface temperature produce variations in the carbon-14 to carbon-12 ratio fractionation at the sea-air interface and hence in the carbon-14 content of atmospheric carbon dioxide and hence in the carbon-14 content of tree rings. But even more effective in perturbing the carbon-14 content of the atmospheric carbon dioxide are the changing magnetic emanations from the sun carrying magnetic fields to the earth in the solar wind. These fields press against the earth's magnetic field, changing the flux of cosmic rays into the earth's atmosphere, thus changing the rate of production of carbon-14 in the atmosphere, produced by cosmic ray collisions with atmospheric nitrogen.

The Bioorganic Reservoir

In the foregoing discussion, we have been concerned with the effects of climate on the biosphere. Now we turn our attention to the effect of the biosphere on climate.

The biosphere is an important reservoir of carbon, derived from atmospheric carbon dioxide, which it both obtains from the atmosphere and stores and returns to the atmosphere as the stored material decays. Its mass depends on the CO_2 content of the atmosphere, on air temperature, on sea surface temperature, namely, on climate, and on the concentration of carbonic anhydrase in the surface seawater. The dependence of rate of absorption of CO_2 into surface seawater has been indicated by Berger and Libby (1970).

In turn, the concentration of CO_2 in the atmosphere depends on the mass of the biosphere and its rate of decay after death and on the carbonic anhydrase concentration in the sea surface. In future predictions of the rate of increase of CO_2 partial pressure in the atmosphere, due to burning fossil fuels, it will be important to include the interaction of the atmospheric CO_2 with the bioorganic reservoir and the catalyzation of its absorption into the sea by means of the action of carbonic anhydrase dissolved in seawater, considerations which have not been taken into account in past computations.

The following data describe the interaction of atmospheric CO_2 with CO_2 dissolved in seawater and with the CO_2 equivalent stored in the bioorganic reservoir:

Total atmospheric CO_2 equals 0.13 g. C/cm.^{2}

Mixed layer of ocean (100-meter-thick surface waters) contains 0.15 g. C/cm.^{2}

0.013 g. C/cm.^{2} exchanged from atmosphere to oceans each year.

0.001 g. C/cm.^{2} returned to atmosphere per year by plant decay.

Storage time in biosphere is about 140 years.

Storage time in oceans is about 1,000 years.

Total depth of oceans of 2,800 meters contains 0.14 g. organic carbon/cm.^{2}

From these data we have shown (Libby 1973) that changes in climate by $\pm 10°$ C can cause changes in the carbon-14 content of bioorganic matter and, therefore, in apparent carbon-14 age of as much as ± 100 years. Such changes are well able to account for Suess' "wiggles" of about 100 years' duration each (Suess 1973) in the carbon-14 concentrations in the 5,000-year sequence of bristlecone pine constructed by Wesley Ferguson. Similarly, the $^{13}CO_2$ concentration of the atmosphere is affected by variations of the biosphere mass, just as is the carbon-14 content, although the amplitude of the perturbation is only half that for carbon-14.

If the stable isotope ratio of $^{13}C/^{12}C$ is to be further measured in tree rings and interpreted as an indicator of climate variation (and we have barely begun to initiate its use as a thermometer in the present work, confining our measurements to the stable isotopes in water, because water is so abundant compared to carbon dioxide and because the dependence of its isotope ratios is relatively simple compared with those of carbon dioxide), certain more sophisticated considerations must be given to the distribution of carbon dioxide among the reservoirs on the surface of the earth.

In the process of making bioorganic matter, plants store carbon depleted with respect to carbon dioxide by about 2.7 percent in carbon-13 (Craig 1963). If the total amount of depleted carbon stored in the biosphere in the past has been different from what it is now, interpretation of $^{13}C/^{12}C$ variations in tree rings as caused solely by temperature changes will reflect an error.

The error results in the following way. The amount of organic carbon dissolved in the oceans is 0.14 g./cm.^{2}, about equal to the amount of carbon in atmospheric carbon dioxide of 0.13 g./cm.^{2} Since the organic carbon is about 2.7 percent depleted in carbon-13, the atmospheric reservoir is correspondingly enriched by about 2.7 percent.

In past times the total mass of stored organic carbon may have been larger or smaller than it is now, depending on the climate. Let us define as the norm the amount of carbon presently stored, M^*, and define a time dependent factor, M/M^*, by which the organic carbon reservoir may be increased ($M/M^* > 1$) as in a lush, tropical coal age or decreased ($M/M^* < 1$) as in an ice age, where M is the mass of organic carbon at the time in the past when the material was alive. We assume that the total amount of atmospheric carbon dioxide has always remained the same (which may or may not be true).

Then the atmospheric reservoir is enriched by about 2.7 percent multiplied by the ratio of the mass of stored organic carbon to its mass today. That is, the enrichment E is given by

$$E = 0.027(M/M^*) \qquad (2)$$

so that at the present time $M/M^* = 1$ and $E_{1979} = E^* = 0.027$ (table 2-2).

Thus the enrichment correction depends on the numerical value of 0.027 depletion of carbon-13 in plants, which in turn depends on the temperature at which the bioorganic material grew. The temperature coefficient for $^{13}C/^{12}C$ in marine plankton has been measured as 0.35 ppt/° C (Sackett et al. 1965) and independently as 0.5 ppt/° C (Eachie 1972). In the absence of more measurements, we may assume it to be the larger of the two experimental values, namely, 0.5 ppt/° C, and this number enters in the calculation of the enrichment, E_t, depending on the average temperature at time t in the past.

Another effect enters also, caused by the presence of ocean carbonate. The sea contains dissolved carbonate that is fractionated in $^{13}C/^{12}C$ in comparison with that in the atmosphere; it is enriched in

Table 2-2. Values of Enrichment of $^{13}C/^{12}C$ Computed for Atmospheric Carbon Dioxide versus Plant Mass M and versus Temperature

M/M^*	10° C	20° C	30° C
0.5	0.015	0.014	0.010
1.0	0.030	0.027	0.020
2.0	0.060	0.054	0.040

Note: This computation is described in Libby 1973; also see appendix 1.

^{13}C and is less so as the temperature increases, causing a small correction to the enrichment *E*.

This perturbation comes only from the carbonate in the approximately 100-meter-thick mixed surface layer; it contains about 0.15 g. carbon/cm.[2] (Kroopnick, Weiss, and Craig 1972), approximately equal to that in the atmosphere and to the organic carbon dissolved in the total depth of the sea. In the case of organic carbon, the total depth of the sea is involved because bacterial decomposition occurs at all levels, producing methane and carbon monoxide, with both of which the sea is saturated, so that these gases are bubbling up from all depths. In comparison, the sea is not saturated with CO_2 at any depth (ibid.).

We have discussed isotope fractionation of CO_2 in the sea surface at the beginning of this chapter.

Other Climate Indicators: Commodities, Prices, and Wages

In order to interpret, verify, and calibrate the temporal changes in the stable isotope ratios of hydrogen and oxygen in tree rings, we look to historic records of crops that humankind has reaped without intensive cultivation, depending instead on the vagaries of climate to produce a good or a bad crop.

The numbers of kinds of major crops have been few, and the records of crop yields have become more reliable only in the last few centuries. We use what records we have been able to find in the libraries of the Department of Agriculture at The Hague, Holland; in the British Department of Agriculture, London; and in the much less lengthy records in the United States.

We analyze here two kinds of records indicative of climate change—the yearly record of wheat prices in four European countries since 1200 and the yield of blue crab in Chesapeake Bay since 1920—as examples of mathematical treatments that could be used on other temporal records of commodities derivable from archives.

In Europe, archivists have pioneered the evaluation of past climate changes from records of wine harvests and from drawings showing the advance and retreat of glaciers. This great body of work is excellently reviewed by E. L. Ladurie (1971). In America, researchers at the Laboratory of Tree Ring Research at the University of Arizona, Tucson, have developed the measurement and analysis of tree ring widths using a score or so of trees in each stand to provide pluviometric maps versus time. They find that these maps have to be established separately for each region and for each continent, so that we cannot look to ring width records versus time for information about the cyclic evolution of climate.

Unlike records of stable isotope variations, records of ring width variations depend on many factors besides temperature alone, for example, on the amount of rainfall and on the amount of volcanic dust fallout, on forest fires, and on animal use of the forest floor, and it is impossible to deduce periodicities from them (Fritts 1966).

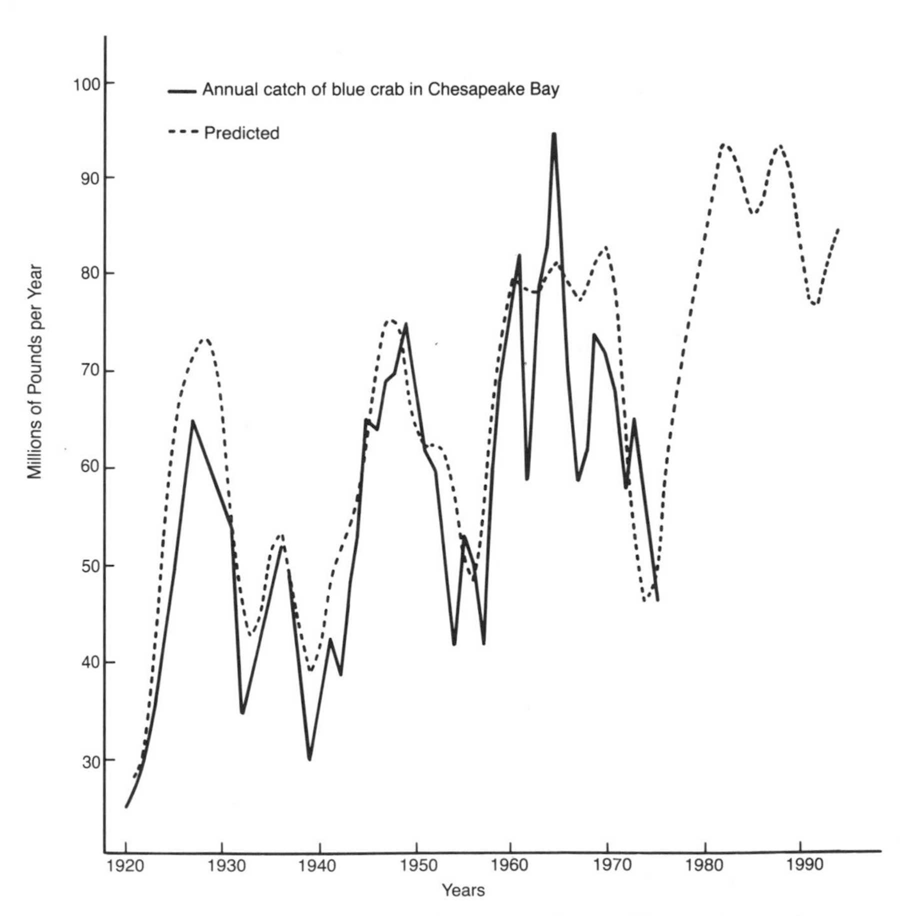

FIG. 2-19. Annual catch of blue crab measured in millions of pounds per year, in Chesapeake Bay, 1920 to 1976.

The blue crab (*Callinectes sapidus*) catches in Chesapeake Bay have been measured (Franklin 1977) for the years 1920 to 1976; see figure 2-19. We have made a Fourier transform of the yield in millions of pounds per year (1 million pounds = 454 metric tons) versus years into amplitude versus period in years—see figure 2-20—and find principal cyclic periods of 8.6, 10.7, and 18 years.

The Fourier transform of the record of annual air temperature at Philadelphia (which is close to Chesapeake Bay) versus time shows principal periods at 8.6, 10.7, and 18 years; the Fourier transform of

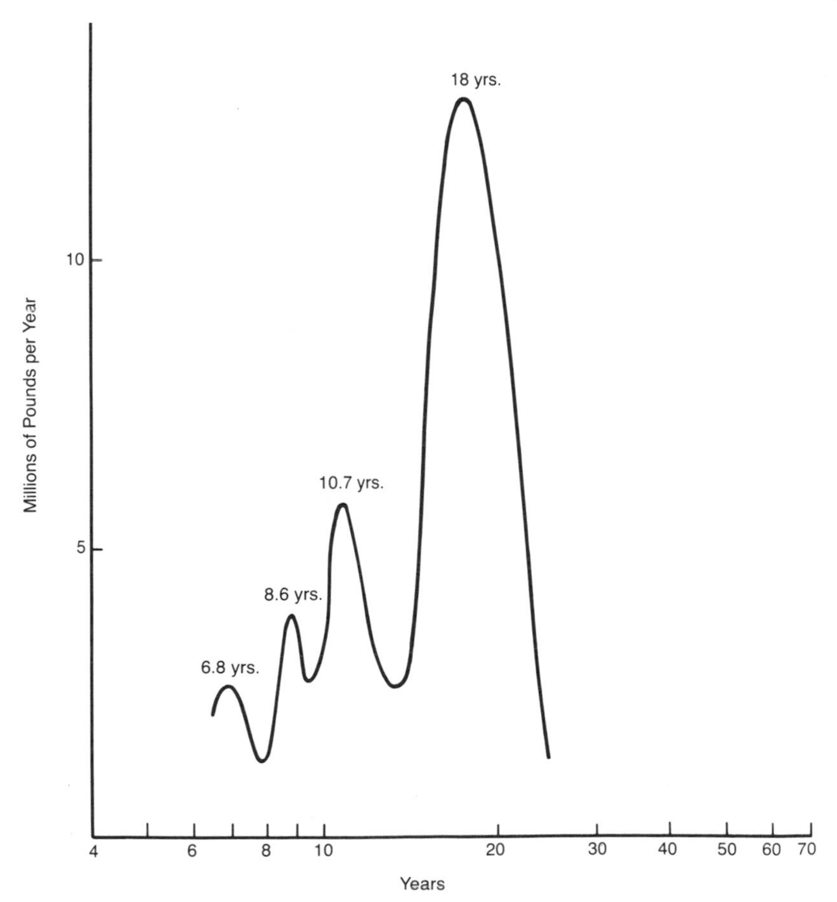

FIG. 2-20. Fourier transform of annual catch of blue crab in Chesapeake Bay versus time, measured year by year, into amplitude versus period.

Table 2-3. Comparison of Periods, Amplitudes, and Phases from Four Natural Phenomena

Annual Average of Philadelphia Temperature			Annual Average of Philadelphia Rainfall (Minima)			Annual Average of Blue Crab Catch in Chesapeake Bay			Earth-Sun-Moon Tidal Forces
τ	ϕ	A	τ	ϕ	A	τ	ϕ	A	τ
17.5	0	1.0	18.5	0	1.0	18.0	0	1.0	18.6
9.8	−1.9	1.2	9.5	−4.1	1.1	10.7	−0.9	0.5	—
7.4	1.4	0.9	8.3	−2.5	0.9	8.6	2.1	0.3	8.8

the record of annual rainfall at Philadelphia shows periods of low rainfall of 8.3, 9.5, and 18.5 years; see table 2-3 (Landsberg 1975; Landsberg and Kaylor 1976).

Tidal forces (enhanced tidal forces lead to high tides) have periods of 8.8 and 18.6 years (Rinehart 1972). The variations in the tidal forces arise because the orbital planes of the moon and the earth are slightly inclined with respect to each other and because the sun, moon, and earth form a mutually rotating gravitational system so that the magnitude of the tidal force depends on their relative positions.

Variations of temperature and rainfall with the sunspot periodicity of 10.5 years are explained by the findings of the CIAP study (Grobecker 1975), in that climate on the ground is affected by variations in the chemical species in the stratosphere whose concentration varies as the sun's spectrum and intensity vary. Why temperature and rainfall near Chesapeake Bay should be affected by variations of the tidal forces is not so clear. However, the atmosphere and stratosphere are pulled away from the earth by tidal forces just as are the waters of the earth. These forces vary by as much as 10 percent during the tidal periods (Rinehart 1972), resulting in density variations in the stratosphere with the same periods; the consequent density variations may affect the relative rates of stratospheric chemical reactions, causing disturbances of temperature and rainfall on the ground with the tidal periodicities.

The periods of fluctuation of blue crab and their absolute phases and amplitudes, as listed in table 2-3, have been used to reconstruct the annual yield of blue crab, Y, and project it into the future (shown by the dashed line in figure 2-19). The prediction formula is

$$Y = \sum_n A_n \sin\left(\frac{2\pi t}{\tau_n} + \phi_n\right) \qquad (3)$$

where ϕ_n is the phase of the nth period τ_n and A_n is its amplitude and where t is time.

The prediction in figure 2-19 shows the recovery of the crab crop already observed in 1977 (Franklin 1977); the yield would increase until 1982 and remain high until 1988, followed by a decrease. We caution that the absolute phases and amplitudes are to be regarded as inaccurate, because their values depend sensitively on the accuracy of the input data. This may be the reason for the disagreements of the relative phases shown in table 2-3 (Hurt et al. 1978).

As with all our Fourier transforms of real data, we test the statistical significance of the periods so revealed by generating appropriate sets of Markovian data, each datum consisting of a constant, a, plus a random number ε_{ij}, where the random number is small compared with a and varies in such a way that the artificial data set so generated ranges between the maximum and minimum of the values of the real data. Each set so generated we subject to Fourier transform, and we look for peaks of amplitude corresponding to those found for the real data. If, in thirty such experiments, we find no comparable peaks, we conclude that the periods found in the true data set have a significance of better than one in thirty, namely, better than 96 percent confidence level. This has been true for all our Fourier transforms of sets of real data until now.

To each data set we make a least squares fit which is then subtracted from the set so that, of the remaining data set, half the data is positive and half is negative, thus removing the "red noise." This subtracted data set is subjected to Fourier transform.

Records of delicacies such as crab extend for less than 100 years. The commodity for which quantitative records of price (contrasted to yield) extend over the longest span, to our knowledge, is European wheat, for which we have found historic prices in England, the Netherlands, France, and northern Italy since 1200 in units of guilders per 100 kilograms (fig. 2-21) (Veenman and Zonen 1938). In all four countries there were manifold increases in prices around 1600 and 1800, caused by something more general than local political and economic maneuvering.

Economists agree that there exists a definite relationship between the market price of a commodity such as wheat and the quantity available (Samuelson 1970). Thus the price of wheat, in historic times of plenty and in times of short supply, should be an indicator of abundance in good climate and of shortage in bad climate and, therefore, an indicator of climatic changes.

Variations in prices of wheat and wages (ability to pay for wheat) from 1200 to the present, compared with variations in average air temperatures from 1650 to the present, enable us to calibrate the modern part of the wage and price record. The population has been increasing during this time span; it is necessary to correct for this as well as for inflation. In figure 2-22 are shown the prices per troy ounce for sterling silver (90 percent pure) and for fine (pure) gold, in English money, since 1200. Sterling prices are more basic than gold prices because sterling was the only official coin of the realm until 1774.

One way of normalizing grain prices to take inflation into ac-

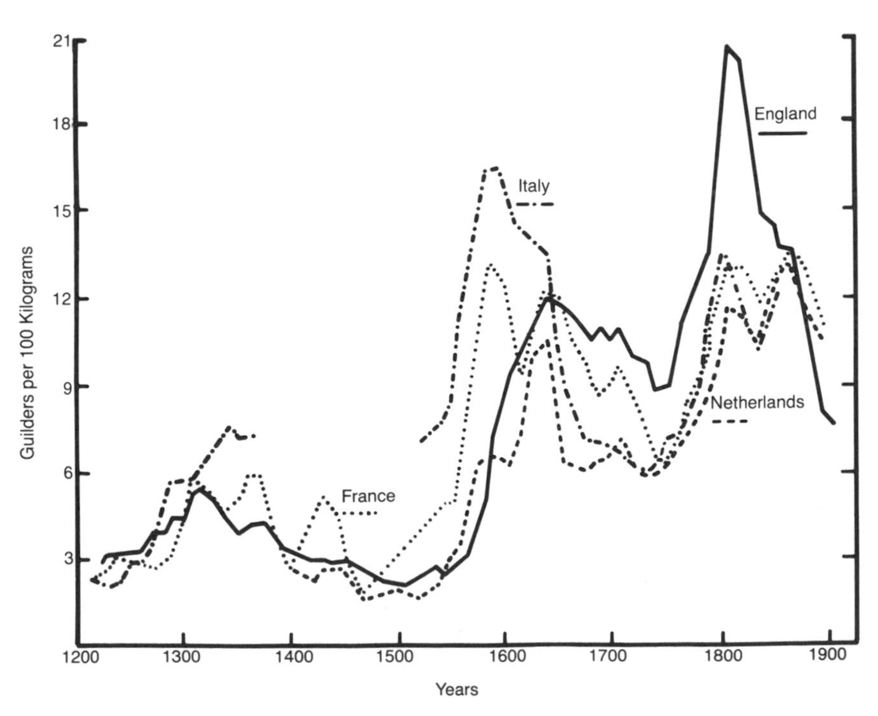

FIG. 2-21. Price of wheat in northern Italy, France, the Netherlands, and England, expressed in guilders per 100 kilograms, versus time, 1200 to 1900.

count is to compute the ratio R_s of wheat prices to the price of sterling, and this we have done. A second way is to compute the amount of grain which a laborer's daily wage can buy, namely, R_w, the ratio of the daily wage to wheat prices. R_s and R_w are shown plotted in figure 2-23 versus centuries. R_s has been computed from data in figures 2-21 and 2-22. R_w has been taken from H. O. Meredith (1939) and G. F. Steffen (1901). The third part of figure 2-23 shows a 25-year running average of winter (January, February, and March) air temperatures in central England computed from G. Manley (1974).

The warming trend between 1800 and 1930 is a general phenomenon displayed in temperature records of Holland, Edinburgh, Stockholm, Vienna, Berlin, Copenhagen, Greenland, the Ukraine, Siberia, Basel, and Geneva; in the United States the trend is evident in New Haven, Philadelphia, St. Paul, St. Louis, and Washington, D.C. (Ladurie 1971). The amplitude of the change varies locally from

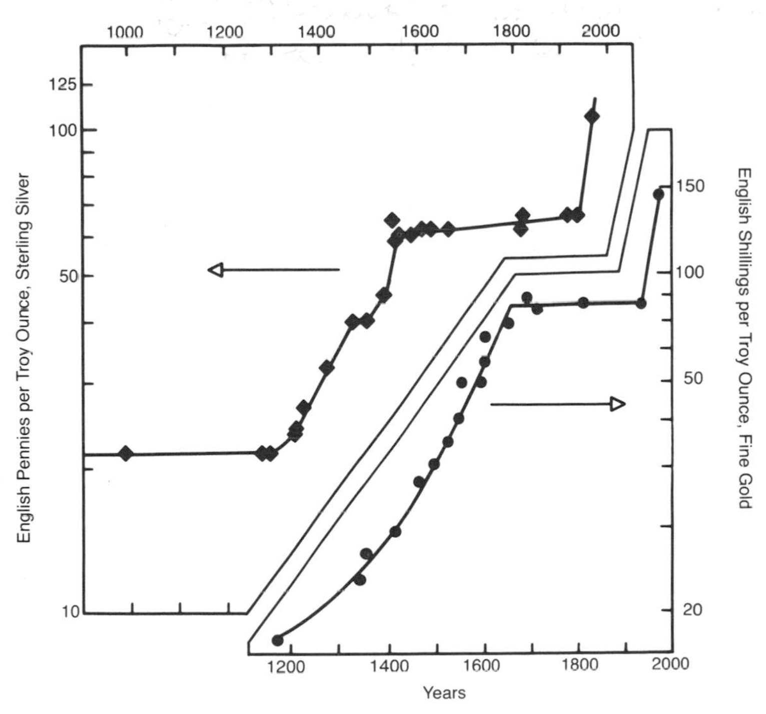

FIG. 2-22. Cost of sterling silver and of fine gold versus time, expressed in English money per troy ounce, 1200 to 1900.

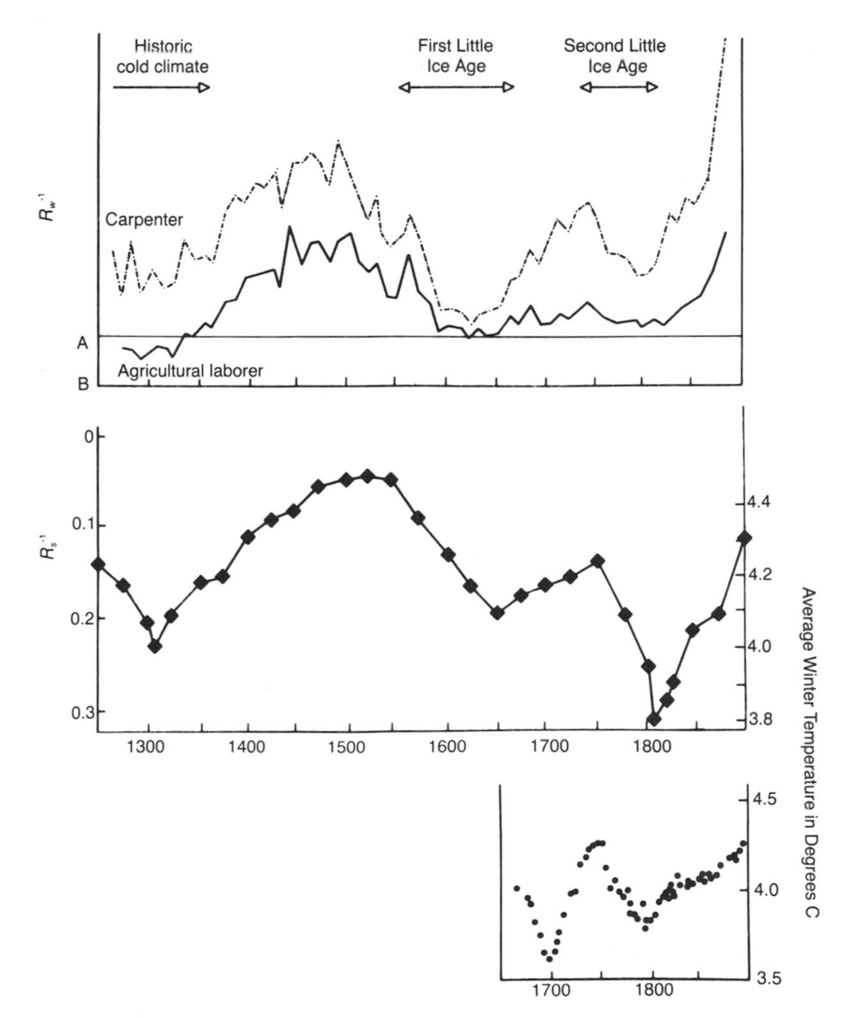

FIG. 2-23. Variations in the amount of wheat purchasable with the daily wages of a carpenter and an agricultural laborer, compared with the cost of wheat and with temperatures measured in central England. A to B represents the quantity of wheat assumed adequate to nourish a family.

1 to 2° C. In particular, the warming of the entire northern temperate zone is estimated at 0.64 to 0.7° C (ibid.) for the period.

The major declines of R_w and R_s around 1630 occurred during the First Little Ice Age, a time famous for famines killing millions of people, cold summers, bitter winters, failed harvests, wars, and civil unrest. The declines of R_w and R_s around 1800 correspond to the Second Little Ice Age, which caused civilian hardships which remain in the political memories of the Irish and Scots today. In Europe in these years there was widespread crop failure and hunger leading to the French Revolution, and in the American colonies there was the revolution against the British crown and hunger and suffering by the American troops during the cold winters.

Our purpose here was to obtain the temperature (T) scale shown in figure 2-23, applicable to times as far back as 1250 when there were no thermometer records, in order to compare the temperature record obtained with the record of tree thermometers. To this end, we measured ΔR_w, ΔR_s, and ΔT (from fig. 2-23), from peaks to valleys, since 1700 and, from these, computed the temperature coefficients $\Delta T/\Delta R_w$ and $\Delta T/\Delta R_s$. Using these coefficients, we computed the temperatures for 1250 to 1600. The resultant temperature scale, shown on the right-hand ordinate of figure 2-23 (Libby 1977), is in agreement with that obtained from variations of the stable isotope ratios in the European oaks.

Let us assume that major variations in climate occur throughout the hemisphere, so that climate declines shown by the Japanese cedar indicate cold times in Europe. Then, in figure 2-9, decreases in the values of the stable isotope ratios indicate cold climate at about 300 and again at about 800 and 1400 to 1500, as well as in the First and Second Little Ice Ages. The cold interval at about 300 coincides with the reign of the Roman Emperor Diocletian from 284 to 305. The interval of cold climate shortly after 800 coincides with crop failures during the reign of Charlemagne. About the cold interval from 1400 to 1500 we have little information, until the reign of the Tudors beginning in 1485, characterized by great monetary inflation and attempts at price and wage fixing such as would be expected as the result of a major climate decline.

In the early part of Diocletian's tenure in office, there began to be frequent invasions of the Roman Empire by barbaric tribes from the north, perhaps motivated by hunger. The emperor increased the armed forces, raised taxes, and debased the Roman coin, the silver denarius. There soon followed a large increase in the money supply, accompanied by soaring prices of commodities and wages of laborers, triggered in part by inflation coupled to failed crops and reduced food supplies. Diocletian attempted to remedy this economic disaster by imposing the famous price and wage controls embodied in the Edict of 301 and by issuing an all-copper denarius which quickly became worthless (Scheuttinger and Butler 1979).

The emperor fixed the maximum prices and wages which could be implemented while allowing it to be legal to charge less and pay less than the maximum, in contrast to the fair-trade laws we have experimented with in the last few years. Parts of the maximum price lists, discovered in about thirty different places, mostly in Greek-speaking portions of the empire, have been assembled and translated by Roland Kent (1920). The edict warns that "if anyone have acted with boldness against the letter of this statute, he shall be subjected to capital punishment," including those who buy and sell at too high a price and those who hoard.

In the edict, maximum prices are listed for grains, dried legumes, wine, oil, vinegar, salt, meats, game, fish and shellfish, fruits and fresh vegetables, milk and cheese and eggs, furs and skins, silk, wool, leather, shoes, lumber, transportation per mile by wagon and by camel, and all other vendible things. Maximum wages were set for farm workers, carpenters, bakers, shipwrights, barbers, teachers, lawyers, and all others who received pay for services.

People stopped bringing supplies to market and instead sold them on the black market, both because they had to sell at prices which paid for their investment and, we suggest here, because a climate decline had reduced the quantities of food for sale. Scarcity produces high prices, as is well known. The working people, unable to buy enough for themselves and their families if they received only the maximum wages listed in the edict, changed jobs, moonlighted, and migrated. Four years after the promulgation of the edict, Diocletian abdicated, citing poor health, and the force of the edict collapsed (Kent 1920; Scheuttinger and Butler 1979).

Charlemagne tried mandating much the same kind of edict during a severe climate decline. Samuel Brittan and Peter Lilley (1977) tell us that he posted tables of maximum prices at times of crop failures and, in an ordinance of 806, stated that "anyone who at the time of the grain harvest or of the vintage stores up grain or wine not from necessity but from greed—for example buying a modius for two denarii and holding it until he can sell it again for four, six, or even more—we consider to be making a dishonest profit." We conjecture that this experiment in price regulation was induced by severe curtailment in food supplies.

That the climate decline in the fifteenth century recorded in our measured Japanese cedar was experienced also in Europe is suggested by historian W. J. Ashley (1923), who tells us that Parliament and the executive left the regulation of prices almost entirely to the local authorities. They fixed prices not only for food but also for other articles needed by the poorer classes, such as wood, coal, tallow, and candles. The evidence of the hydrogen and oxygen stable isotope ratios in the cedar also records climate declines coincident with Diocletian's edict and Charlemagne's price-fixing.

Thus the specific evidence of the stable isotope ratios in the cedar, as well as the coincidence of climate periods in records of the cedar, the Greenland ice cap, the bristlecone pines of southern California, and the sea core from the Santa Barbara Channel, indicates that climate changes occur throughout the hemisphere and are caused by variations in the ultraviolet emissions of the sun.

Additional evidence for the simultaneous variation of climate at widespread places in the northern hemisphere has been found by the historian Samuel Eddy (1980). He notes that thick rings in trees in California coincide with thick varves in Lake Saki in the Crimea, with historic floods of the Nile, and with abundant wheat harvests of classical Greece and Rome, indicating that years of abundant rain were simultaneous on both sides of the earth.

References

Ashley, W. J., 1923, *An Introduction to English Economic History and Theory*, Longmans, London.

Becker, B., and V. G. Siebenlist, 1970, Flora *159*, 310–346.

Berger, R., and W. F. Libby, 1970, in *Global Effects of Environmental Pollution*, ed. S. F. Singer, D. Reidel Publishing Co.

Bergthorsson, P., 1962, *Proc. Conf. Climate 11th to 16th Centuries*, Aspen, Colorado, National Center for Atmospheric Research, Air Force Cambridge Research Laboratories.

Blackman, B. B., and S. W. Tukey, 1958, *The Measurement of Power Spectra*, Dover, New York.

Brittan, S., and P. Lilley, 1977, *The Delusion of Income Policy*, Temple Smith, London.

Chu, Ko-Chen, 1973, Scient. Sin. *16*, 226–250.

Cohen, T. J., and P. R. Lintz, 1974, Nature *250*, 398–399.

Cohn, M., and H. C. Urey, 1938, J. American Chem. Soc. *60*, 679–687.

Craig, H., 1963, Consiglio Nazionale delle Recherche, Spoleto, Italy, Sept., 17–53.

Dansgaard, W., S. J. Johnson, H. B. Clausen, and H. J. Langway, 1971, Nature *227*, 482–483.
DARPA (Defense Advanced Research Projects Agency), 1971, Contract Nos. F44620-73-0025, F44620-72-0029, monitored by Air Force Office of Scientific Research, Washington, D.C.
Daugherty, G. V., 1976, acting superintendent, United States Dept. of the Interior, National Park Service, Three Rivers, Calif., collected the *Sequoia gigantea* and sent it to us.
Degans, E. T., R. R. L. Guillard, W. M. Sackett, and J. A. Hellebust, 1968, Deep Sea Res. *15*, 1–9.
Doose, P. R., 1978, Ph.D. dissertation, Geochemistry Dept., University of California at Los Angeles.
Eachie, B. S., 1972, EOS Trans. AGU. *53*, 406.
Eddy, S. K., 1980, Syracuse Scholar *1*.
Epstein, S., and C. J. Yapp, 1976, Earth & Planet. Sci. Lett. *30*, 252–261.
Ergin, M., D. Harkness, and A. Walton, 1970, Radiocarbon *12*, 495.
Franklin, B. A., 1977, New York Times, June 27, p. 14.
Fritts, H. C., 1966, Science *154*, 973–979.
Gould, R. F., ed., 1966, *Lignin Structure and Reactions*, Advances in Chemistry Series 59, American Chem. Soc., Washington, D.C.
Grobecker, A. J., ed. in chief, 1975, *Monograph No. 1*, Climatic Impact Assessment Program, Dept. of Transportation, Washington, D.C.
Herzberg, G., 1945, *Infrared and Raman Spectra of Polyatomic Molecules*, Van Nostrand & Reinhold, New York.
Hill, J. R., 1977, Nature *266*, 151–153.
Huber, B., 1935, Ber. at. Bot. Gesell. *53*, 711–719.
Huber, von K. M. B., and V. G. Siebenlist, 1969, Sitzberichte Abteilung I. Österreichische Akademie der Wissenschaft *178*, 37–42, Vienna.
Hurt, P. H., 1978, master's thesis, Chemistry Dept., University of California at Los Angeles.
Hurt, P. H., L. M. Libby, L. J. Pandolfi, L. H. Levine, and W. A. Van Engel, 1978, 1979, Climatic Change *2*, 75–78.
Hurt, P. H., L. J. Pandolfi, and L. M. Libby, 1978, detailed in L. M. Libby and L. J. Pandolfi, 1980, Environment International *2*, 317–334.
IAEA (International Atomic Energy Agency), 1969–1975, *Environmental Isotope Data, Nos. 1–5*, Vienna.
Kalil, E. K., and I. R. Kaplan, 1976, Ph.D. dissertation, Geochemistry Dept., University of California at Los Angeles.
Keeling, C. D., 1960, Tellus *12*, 200–203.
Kent, R. G., 1920, Univ. of Pennsylvania Law Rev., 35–47.
Kroopnick, P., R. F. Weiss, and H. Craig, 1972, Earth & Planet. Sci. Lett. *16*, 103–110.
Ladurie, E. L., 1971, *Times of Feast; Times of Famine*, Doubleday, Garden City, N.Y.
Landsberg, H., 1975, *Time Series of Temperatures and Rainfall from Rec-*

ords in the Eastern U.S., Reduced to Philadelphia, Univ. of Maryland, College Park.

Landsberg, W. H., and R. E. Kaylor, 1976, J. Interdiscip. Cycle Res. *7*, 237–243.

Libby, L. M., 1972, J. Geophys. Res. *77*, 4310–4317.

Libby, L. M., 1973, J. Geophys. Res. *78*, 7667–7670.

Libby, L. M., 1974, *Final Technical Report on Historical Climatology*, DARPA order no. 1964-1, Air Force Office of Scientific Research, Washington, D.C.

Libby, L. M., 1977, Indian J. Meteorol. *28*, Apr.

Libby, L. M., and L. J. Pandolfi, 1973a, pp. 21–39 in *Proceedings of International CLIMAP Conference*, Norwich, Eng.

Libby, L. M., and L. J. Pandolfi, 1973b, Contribution 219, *Colloques Internationaux du Centre National de la Recherche Scientifique*, Gif-sur-Yvette, France.

Libby, L. M., and L. J. Pandolfi, 1974, Proc. Nat. Acad. Sci. *71*, 2482–2486.

Libby, L. M., and L. J. Pandolfi, 1976, J. Geophys. Res. *81*, 6377–6381.

Libby, L. M., and L. J. Pandolfi, 1977, Nature *266*, 415–417.

Libby, L. M., L. J. Pandolfi, P. N. Payton, J. Marshall III, B. Becker, and V. G. Siebenlist, 1976, Nature *261*, 284–288, and references therein.

Manley, G., 1953, Quart. J. Roy. Soc. *79*, 242–261.

Manley, G., 1959, Archiv. meteorol. geophis. Bioklmatol. *9*, 3/4, Vienna.

Manley, G., 1974, J. Roy. Meteorol. Soc. *100*, 389–405.

Meredith, H. O., 1939, *Outline of the Economic History of England*, Sir Isaac Pitman & Sons, London.

Michael, H., 1976, University Museum, University of Pennsylvania, counted the rings and dated them by comparison with the fiducial ring sequential pattern developed at the University of Arizona.

Olsson, I. U., and M. Klasson, 1970, Radiocarbon *12*, 281–284.

Olsson, I. U., and A. Stenberg, 1967, pp. 69–78 in *Radioactive Dating and Methods of Low Level Counting*, IAEA Symposium, Monaco, Mar. 2–10.

Pandolfi, L. J., E. K. Kalil, P. R. Doose, L. H. Levine, and L. M. Libby, 1978, pp. 740–755 in *10th International Radiocarbon Conference*, Bern & Heidelberg, Aug.

Rebello, A., 1975, in *Environmental Biogeochemistry*, ed. J. Nriagu, 2 vols., Ann Arbor Science Publishers, Ann Arbor.

Rinehart, J. S., 1972, Science *177*, 346–347.

Rittenberg, D., and L. Pontecorvo, 1956, Intl. J. Appl. Radiat. Isotopes *1*, 208–214.

Sackett, W. M., W. R. Eckelmann, M. L. Bender, and A. W. H. Be, 1965, Science *148*, 235–237.

Samuelson, P. A., 1970, *Economics*, 8th ed., McGraw-Hill, New York.

Scheuttinger, R. L., and E. F. Butler, 1979, *Forty Centuries of Wage and Price Controls*, Heritage Found., Washington, D.C.

Sepall, O., and S. G. Mason, 1961, Can. J. Chem. *39*, 1934–1943.

Soutar, A., and J. D. Isaacs, 1969, *State of California Marine Research Committee, California Cooperative Fisheries Investigation 13*, 63–70.

Steffen, G. F., 1901, *Studien zur Geschichte der englischen Lohnarbeiter mit besonderer Berüchsichtigung der Veränderungen Ihrer Lebenshältung*, 3 vols., Hobbing & Buchle, Stuttgart.

Suess, H. E., 1973, Endeavor *32*, 34–38.

Taube, H., 1956, Ann. Rev. Nucl. Sci. *6*, 277–302.

Veenman, H., and N. Zonen, 1938, *De Landbouw in Brabants Westhoek in Het Midden van de Achttiende Eeuw*, Agronomisch Historische Bijdragen, Wageningen, Netherlands.

Vogel, J. C., and J. C. Lerman, 1969, Radiocarbon *11*, 385.

Waldmeier, M., 1961, *The Sunspot Activity in the Years 1610–1960*, Schulthen & Co., Zurich.

Weast, R. C., ed., 1962, *Handbook of Chemistry and Physics*, 45th ed., Chem. Rubber Co., Cleveland.

World Weather Records, 1966, vol. 2: *Europe*, U.S. Dept. of Commerce, Environmental Services Administration, Washington, D.C.

Yamamoto, T., 1971, Geophys. Mag. *35*, 187–206.

3. Human Interaction with Climate

The Relation of Tree Thermometry to Meteorology and Geophysical Climate Assessment

As our ability to evaluate historic and prehistoric climates and to predict the immediate and near future changes in climate improves, we begin to appreciate the importance of such an ability to future planning on the federal and international level. The plans for national defense will become increasingly dependent on the futures of food production, transportation, the effect of land, sea, and air disturbances on communications, energy production (especially electricity production), and the demands of other, less technologically capable countries to share what we have developed.

Food production, distribution, and consumption are among the most energy-consuming of human activities. Today we are farming with natural gas and oil. For example, several grams of crude oil or its energy equivalent are required to produce each gram of dried corn. The energy inputs include ammonia fertilizer, fuel for tractors, trucks, and harvesting equipment, fuel for drying, and fuel for transportation to storage and to market. Additional increments of energy are needed to process agricultural products for human consumption. Besides these obvious inputs, there are others—such as energy for cold storage and freezing, nuclear irradiation for sterilization and pest control, energy for pesticides and mold inhibition, chemicals to speed ripening, and the like.

Animal wastes cannot compete with organic nitrates in effectiveness as fertilizers and are far more expensive to distribute. The production of nitrate fertilizers requires ammonia, which is derived from natural gas. As natural gas becomes more expensive, hydrogen for the manufacture of ammonia will have to come from some other source, such as electrolysis of water.

As world food supplies become increasingly inadequate, the allocation of fuel for food production takes on an increasingly higher priority, thus competing with fuel allocations for electric power production, transportation, communications, and so forth. The reliable prediction of competing demands for primary fuels is fundamental to planning for the electric power industry, in particular. The inevitable and anticipated shortage, in the future, of fuels and minerals requires us to seek to identify relationships between climate variations and energy production, with the objective of finding trade-offs that can lead to more effective protection of humanity from climate deterioration.

World food supplies have now, in the early 1980s, reached a low of enough for only a few dozen days. The climate of the northern hemisphere appears to be deteriorating rapidly, thus threatening to reduce these reserves even more. Each year there are ominous predictions of crop failures in countries of high northern latitudes. In any years that there would be a simultaneous failure of grain crops in many countries, the United States would be under severe pressure to ship grain gratuitously in amounts far larger than has ever been the case in the past. Obviously, in planning our agricultural program, we need to know as much as we can about the weather trends in the near future. These matters are of worldwide concern but are particularly important to the United States, because the great North American landmass and our agricultural technology have become increasingly important to feeding the world. This will place an increasing load on demands for oil and natural gas, and on their prices, and will create competition for other users of massive amounts of fuel (Brown 1981; Barr 1981).

The use of energy in the preparation and delivery of food depends in part on dietary customs and the ways they are changing. In the last year for which fairly complete records are available, 1963, the U.S. food cycle is estimated to have used 12 percent of the national energy production and 22 percent of the total kilowatt-hour consumption. About half of the 22 percent of the electricity for food production was used for food preparation in homes, and it should be noted that in electricity production half of the energy is thrown away as waste heat which could have been used for food preparation, and use of electricity for cooking throws away another half (Cervinka et al. 1974), namely, half of that remaining.

The trend in 1963 appears to be that the food-related energy use was growing faster than the population. To alter this trend, it follows that the eating habits of the nation will have to change away from prepared foods, food additives, and foods of no nutritional content and that cooking habits will have to change back toward quick techniques with hot fires that consume little fuel, such as were used by prehistoric people for the last million years and such as are used today in fuel-poor countries like India and China.

Alvin Toffler (1980) has identified three examples of the huge and relatively sudden perturbations in the human way of life caused by climate changes in the recent past. The "First Wave" of change was triggered some 10,000 years ago by the invention of agriculture. Obviously this invention followed on the retreat of the continental glacier which covered the British Isles, Europe, and Asia, accompanied by the climate becoming more clement.

Toffler's "Second Wave," launched over two centuries ago as the Industrial Revolution, began during the Little Ice Age, when the climate suddenly and briefly changed very much for the worse. The consequent widespread starvation and reduction in standard of living led millions of people to live out their dreams through their children. They hoped and expected that their children would do better socially and economically, and indeed, as the climate again improved, their hopes and expectations were realized.

The "Second Wave" was accompanied by the spread of the Protestant ethic, with its emphasis on thrift, toil, deferral of gratifications, and channeling of energy and earnings into economic development. Government became aware of the need for defense against the vagarics of climate and began to fund research to improve agriculture and to develop methods of food preservation that would keep food wholesome for periods longer than a year, which had already been achieved millennia ago by simply drying it.

Research on better and new kinds of crops and intensive agriculture now, after over two centuries, is of course a well-funded and productive science. What we have not yet achieved is a proven method of predicting climate changes in the near and not so near future which would perturb food production, transportation, public health, communications, fuel supplies, per capita consumption of electricity, and the like.

The importance of the prediction of national and global food production to planning in the U.S. Department of Defense is obvious. Equally obvious is the department's need to anticipate disturbances of the solar surface, which in turn cause perturbations of the earth's ionosphere, inhibiting transmission of communications in the atmosphere and ionosphere, in fact, even in underground cables; these perturbations also cause misbehaviors and interruptions of the electrical utilities industry and changes in the sea surface temperature affecting climate.

Predictions of future solar fluctuations until now have been made using the record of sunspot variations of more than 300 years, almost the only solar record we have. There is, of course, the record of the last 30 years or so of the ionospheric, auroral, and solar coronagraph observations, but these are of a duration too short to be useful yet in predictions extending several years or tens of years into the future.

The auroral record has been compiled from six auroral catalogs covering the period 5000 B.C. to A.D. 17,000, combining both Oriental and European observatories (Siscoe 1979). The combined catalogs provide a sufficient number of events to demonstrate an 11-year (probably sunspot) periodicity and an 80-year periodicity. An important deduction from the combined record is that these forms of solar variations have operated for at least the last 7,000 years.

Yet there is general agreement, based on the biological record, that the solar energy output has practically not changed in the last 3 billion years. But it is known, from balloon and satellite measurements, as we have said earlier in this volume, that the ultraviolet solar energy fluctuates by factors of 2, 3, and more. Thus, aside from parameters important in climatic change, such as the precession of the earth about its axis of rotation and changes in the sun-earth distance and planetary alignments, the future changes in the solar ultraviolet flux and their effect on climate can be predicted only if we know its past history.

The method of tree thermometry, in principle able to be extended into the past as far back as trees of those times which still remain today can be found (perhaps for the last 50,000 years), alone offers itself as able to evaluate the past history of those times and thus may predict some short span of the future climate.

Tree thermometers allow us to evaluate climate periods with an accuracy of 1 year. Thus, the method of tree thermometry fits between the method of present-day computerized meteorology, which is able to predict climate for perhaps a few months into the future by an extrapolation of the climate of today, and the many geologic methods of assessment of past climates, which have accuracies of no better than 1,000 years or more and thus are able to predict climate periods of more than 1,000 years but not less.

According to meteorologist Jerome Namias (1980), rigid statistical analysis, taking 50 percent as chance forecasting, has achieved something like 65 percent to 70 percent correct monthly and seasonal forecasting. The reason for this low level of accuracy is that a hemispheric or global data base of uniform goodness is necessary, whereas there are large areas of the world for which no meteorological data are gathered. But, even if the data base were much better, forecasting probably would not improve much because there is no physical understanding of the causes.

Forecasters use the simple idea that the weather repeats itself. They select cases in the past that resemble the present. They read off what happened next in the past and use that as the prediction for what will happen next in the present. Unfortunately, only about 30 years of historical surface and upper air weather data are available from which to select matching events. Furthermore, this analogue method of forecasting employs neither understanding nor imagination. Unless forecasters can explain how and why trends develop, their extrapolations of weather changes for even a few days ahead are likely to be wrong. Also, even with perfect data and computers of infinite capacity to store data, computerized predictions of weather for 2 weeks ahead become overwhelmed by round-off errors on the ends of the numbers which the computer is churning out (Namias 1980).

The present-day computerized atmospheric models of climate may be represented by the rather simple model used at the Rand Corporation (Gates et al. 1971; Gates and Schlesinger 1972; Gates 1976). All such models represent the global tropospheric circulations with mathematical equations for large-scale horizontal atmospheric motions subject to the conservation of energy, the hydrostatic equations requiring air to flow from regions of high pressure to regions of low pressure, and the continuity equations for air mass and mass of water vapor (mass is conserved). These equations are bounded by such requirements as a fixed temperature for the surface of the sea, another fixed temperature for the surface of ice caps, specified albedos (reflectivities) of sea, air, ice, and land, and specified areas of sea, air, ice, and land.

The equations of the Rand model are written in terms of two layers of the troposphere of equal mass, between the surface air pressure (variable) and the assigned upper 200-millibar surface. Solutions of the model's equations, specified on a grid on the spherical earth at points separated by 4° in latitude and 5° in longitude, are computed by stepwise numerical integration for every 6 minutes. Computed variables include ground temperature and pressure, humidity, and precipitation at each point on the grid. Fifteen minutes of computer time simulate 1 day's change. Thus 2 days of computer time are needed to predict climate 6 months into the future.

The methods of geologic investigations to assess past climates use inferences from a variety of phenomena. Oil deposits, reefs, corals, and organic limestones give evidence for the previous existence of warm seas. Coal beds testify to swampy, warm, wet tropics with high rainfall. Salt and other evaporite deposits, such as gypsum, halite, and potash, indicate climates in which the evaporation rates were high over bodies of salt or brackish water. Red beds (containing Fe_2O_3) indicate arid desert conditions during their formation. Red beds are absent from areas which had excessive or even moderate rainfall because there the brown iron oxides formed instead.

Wind directions deduced from scratched rock and from ripples hardened into rock give some idea of monsoons, trade winds, and other prevailing winds. Laterite deposits, some of which show seasonal variations or varves, made for the most part of Al_2O_3 and Fe_2O_3, may be produced by wind or water; these develop in tropical savannas and as a result of advances and retreats of glaciers producing tillite moraines.

Interbedding of seams of coal with sandstone and shales indicates cyclic behavior of past climates, in which glacial and nonglacial intervals succeeded each other at commensurate intervals. Sedimentary layers deposited in more recent times contain pollens of various plant species, indicating warm, cool, wet, or dry growing conditions, and show cyclic successions of such conditions by interbanding of successive layers of differing pollen content.

Analysis of the ocean bottom cores for variations in the stable isotope ratio of $^{18}O/^{16}O$ in shells of foraminifers which grew in the ocean surface and of $^{13}C/^{12}C$ in bottom organic sediments, formed by bioorganic material which grew in the ocean surface and precipitated after death, allows the temperature of those sea surfaces to be deduced as a function of time back for about 100 million years, the maximum age of ocean bottom sediments.

At least three times in the history of the world all the continents have been subjected to ice ages: 600 million years ago, 275 million years ago, and the ice age which began about 1 million years ago and from which we seem to be emerging. Caution is indicated in making this statement, because polar ice caps still exist and did not exist in previous warm intervals. In the last 100,000 years, the northern hemisphere (at least, and probably all the continents) has been in an ice age 90 percent of the time (Emiliani 1978).

The Quaternary ice age in the northern hemisphere, foreshadowed by events during the Tertiary period, remains one of the most astonishing climatic events in the history of the earth, involving a marked lowering of temperatures and the repeated succession of glacial and interglacial epochs, each relatively short. In contrast, the late Paleozoic glaciations play only an insignificant and indirect role in the history of the northern hemisphere. Of the significance of the early Cambrian glaciations little is known (Schwarzbach 1961).

In Europe and North America, during the last 500 million years, the average climate was warmer and drier than it is today. At the beginning and end of this interval there were ice ages with extensive glaciers, those of early Cambrian and of the Quaternary (the last million years). During this time, however, there was a gradual cooling as evidenced by the shift of the coral reef zones from polar regions to the present equatorial positions, and there was a similar movement of the evaporite belts. But one must remember that, in 500 million years, continental drift (some 1 to 2 centimeters per year) was great enough to open up the Atlantic Ocean, to move Antarctica to the South Pole, and to move Europe and Asia to their present positions at quite northerly latitudes.

Up to the beginning of the Tertiary period, about 70 million years ago or so, the picture of the earth's climate can be reconstructed only in the broadest outlines. However, in the Tertiary the plants and animals began to correspond closely to modern forms and so became the most important of climate indicators. Throughout the Tertiary the climate was warmer than at present, but it cooled slowly until, by about 1 or 2 million years ago, it had become about the same as at present (Schwarzbach 1963). And about then one finds the first conclusive evidence for human beings (Howell 1972, e.g.), the date being determined by potassium-argon dating.

The principle of potassium-argon (^{40}K-^{40}Ar) dating is that, when sediments are being deposited, the naturally occurring radioactive potassium isotope ^{40}K is contained in the soil grains, but its daughter of decay, the isotope ^{40}Ar of the rare gas argon, completely diffuses out of the grains because they are small. After deposit argon can no longer escape. Then, using the half-life of the radioactive decay of ^{40}K (1.28×10^9 years), and assuming that neither parent nor daughter has been lost from the sediment since its deposit and that all other argon in the sediment has the isotopic composition of present-day atmospheric argon, one can compute the time since the sediment was deposited from the ratio of $^{40}Ar/^{40}K$. This dating method can be extended down to as low as 100,000 years, but with increasing difficulty and unreliability (Fitch 1972).

At ages of less than 100,000 years and especially more certain at ages of less than 50,000 years, the dating method of radioactive carbon-14 becomes viable. The limiting factors are paucity of decays in very old samples and the likelihood of contamination by more modern dirt, tree roots, and the like.

The climate assessment technology of tree thermometry will be able to make an important contribution to history from the present back to 50,000 years and perhaps to 100,000 years, as far back as old trees can be found, so it is this period of time which interests us here. In particular, it is in this period that hominids evolved into modern *Homo sapiens* and developed their technological competence with increasing rapidity through the ice ages to achieve a base on which their capabilities have blossomed into the historical era, with its salubrious climate compared with that of the last ice age.

We may ask the method of tree thermometers to give us an understanding of how climate change has acted to enable humans to achieve this remarkable technological evolution so recently, in only a few scores of thousands of years, when they had remained culturally almost static for 2 million years. The answer may be partly genetic: *Homo sapiens sapiens* were able to achieve greater orders of magnitude in invention and development than could their predecessors because they benefited from a genetic mutation.

But, insofar as *Homo sapiens sapiens* have learned how to mitigate the effects of climate and survive climate change, we need to study how climate may have affected their progress with the invention and development of technology. We want to evaluate the rapidity of improvement of their ability to dominate natural forces in recent millennia compared with all the time that went before. For example, we may ask, did alternation of glaciation with intervals of good climate stimulate technological development?

Early Hominid Evolution in Changing Climates

We describe here the achievements of the earliest known hominids, of a few million years ago, those to be found in East Africa, in Ethiopia, Kenya, Jordan, and Tanzania, and in South Africa (Johanson and White 1979; Johanson 1976, 1978) and most recently in Nepal (Munthe 1981) and Pakistan (Pilbeam 1982). It should be mentioned that there are two earlier hominid fossils, at 14 million years and at 5 million years, but without significant multiplicity and without artifacts (Bishop 1971).

For the most part, dozens of skeletons 1 to 4 million years old have been found along the Great Rift Valley as it traverses East Africa from Ethiopia to south of Tanzania (fig. 3-1). Some, of about 2 million years ago, have been found in the Transvaal of South Africa. In Ethiopia along the Awash River an almost complete 3-million-year-old skeleton of an adult female has been found; the individual was about 4 feet tall and capable of walking upright. Altogether a remarkable collection of hominid skeletons, representing from thirty-five to sixty-five individuals, has been recovered (Johanson and White 1979). Remains of fossil animals, such as turtles, crocodile eggs, and crabs, in the same strata indicate that the hominids lived along a lake edge in lush grass and woodlands, and the fossil animal remains may indicate what they ate. Many of the bones were found all in the same place, perhaps a graveyard, and represented male and female adults and children. No artifacts were found with them, and no living sites were found.

The environment of 1 to 5 million years ago along the Rift Valley, where 2-million-year-old hominids and their living sites and artifacts have been found, was characterized by volcanic highlands

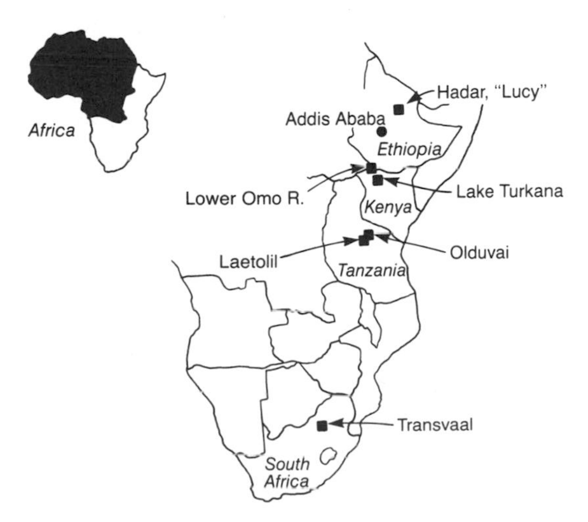

FIG. 3-1. Major sites of hominid homes, 1 to 4 million years old.

transected by deep grabens, diverted and reversed rivers, perturbed as the result of being dammed with volcanic ash, and volcanoes with forested slopes and radiating streams. This environment undoubtedly supported a rich and nutritious plant and animal life able to feed hominid families and small tribes (Shackleton 1978), who left their skeletons, stone tools, and food residues as evidence of their lengthy and flourishing occupation.

Subsequently, *Homo erectus* spread into many parts of Europe and Asia until as late as half a million years ago. Then the basic *Homo sapiens* emerged, to be followed by *Homo sapiens neanderthalensis*, some 150,000 years ago, and finally by *Homo sapiens sapiens* 40,000 to 50,000 years ago.

From the first dates of the *Homo* skeletons, at least 2.5 million years ago, there is evidence for ancient campsites where stone tools were made; these campsites are littered by the shards of stone raw materials and by garbage heaps containing evidence of meat eating (plant residue and wooden tools have long since disintegrated). In the period from 3 million years down to about 0.5 million years ago, when *Homo* sites spread to Europe and Asia, there was a steady increase in the number of camps and in the complexity of the tool technology discarded in them.

By 1.5 million years ago, the basic kit of Olduvai, containing six kinds of stone tools, had expanded to ten types of implements. Implements were easily made and expendable, often worn out and discarded at the site of a kill it would seem—as witness bed IV, site H.K., at Olduvai, which yielded 459 hand axes and cleavers blunted by use lying amid the disarticulated skeleton of a hippopotamus (Clark 1969).

Mary Leakey discovered, excavated, and examined the first substantial samples of very early stone tools from a stratified context of about 1.6 to 1.8 million years old. She noted that hominid stone breaking by percussion was effected to produce small flakes with sharp edges and pointed angles that could be used to skin and cut up animals. These tools were struck from larger rocks, called cores, which both formed core tools and served as the source of more little knife blades, which were banged off as needed. Many artifact-bearing camp and manufacturing sites have been discovered in the neighborhood since then, of ages 3 to 1 million years B.P. In fact, Olduvai and its surroundings have provided the longest known record of the stone tool industry, spanning a period of about 3 million years (see fig. 3-1).

The earliest of the living and manufacturing sites, characterized by abundant choppers and small cutting tools, are mainly along the shores of the former Olduvai Lake. The hand ax culture appears about 1.4 million years ago. The tool industry in the later stages shows an expanded kit with a greater variety of instruments. All dates are measured using the K-Ar method (Isaac 1978). There is clear evidence for efficient hunting in the heaps of animal bones at the campsites. Thus the tool-making industry and the hunting, slaughtering, and butchering industry had already been initiated (Harris 1981).

The tools include (1) core tools for heavy-duty functions, such as choppers, discoids, protobifaces, and polyhedrons, (2) small scrapers, awls, burins, and chisels, (3) flakes and splinters for knives, and (4) hammers and anvils. The choppers are usually made of lava, whereas the flakes are mainly of quartz, fracturing of which produces glassy chips with very sharp edges. The core rocks were brought by hominids to the manufacturing sites and campsites over many kilometers, up to 100 kilometers away.

These same skills are still known today, as was evidenced by an incident reported by Isaac: his group, accompanied by a young local boy, came on a large animal kill freshly abandoned by a lion. The boy banged off a chip from a nearby rock, with which he skinned an untouched leg and carried it along for the evening meal.

Stone Age skills, in fact, were practiced into this century in North America by Indians. At the time that the whites first encountered American Indians in the Far West, Indians were Stone Age people. Their ingenuity had developed, for example, the taking of fish by a multiplicity of methods, including dip nets and seines, snares or nooses, spears, harpoons, bows and arrows, herring rakes, weirs and other traps, and poison.

Fishing with a noose is described by Fools Crow (ceremonial chief of the Teton Sioux, born in 1890), who was still of the Stone Age culture. "There were lots of bass in the creeks. I went fishing for them quite a bit. My stepmother taught me how to do this, with a loop made from the white braided hair of a horse's tail. I would lay on the bank and watch the schools of fish. The bass have a habit of going through things so I maneuvered the loop in front of them, and then as one went through it I pulled it tight to catch him." Of the bow and arrow he recalls, "When I was nine or ten years old, my grandfather made me a bow and some arrows and taught me how to use them. I began to hunt rabbits, prairie dogs and prairie chickens, and in time I became an excellent marksman, bringing a considerable amount of game home to add to our food supply" (Mails 1979).

He was probably the last of people dependent on the bow and arrow for meat. Except for the bow and arrow and the canoe, his skills and needs would probably have fitted him very well for coexistence with million-year-ago hominids. Similarly, the Batete people of Botswana have tribal memories extending back 100 to 300 years ago of how they hunted hippos by harpooning them with spears from safe positions in trees (Rubin 1981).

Hominid life of about a million years ago, reconstructed from the evidence of the artifacts, living floors, and bone piles, was spent in relatively open habitat (perhaps with only insubstantial, readily movable shelters). Bipedal upright locomotion was used, as proven by the hominid footprints found at Laetolil, which were made in soft mud and later hardened into a permanent record (Leakey 1979). People ate meat cut from carcasses as large as or larger than themselves and they carried meat back to the base camp, where very probably both meat and gathered plants were shared. The most recent findings indicate that these people were able to use fire and cook their food as far back as nearly 2 million years ago (Brain 1981).

The fossil human footprints found at Laetolil, about 20 miles south of Olduvai Gorge in northern Tanzania, were at the edge of a former water hole, in use 3.5 million years ago. There was a profusion of animal and bird life, which slaked its thirst there and left petrified remains in 500 feet of dry deposits—giraffes, rhinos, dinotheres (elephantlike creatures with tusks protruding from their lower jaw rather than growing from the upper jaw, and flanking the trunk), monkeys, rodents, hyenas, and many other different types of animals, most of which are now extinct. Besides human footprints, there are many footprints of these animals (Leakey 1978).

For all modern humans, the length of the foot is about 14 percent of the height. From the footprints in Tanzania and Ethiopia, using the same rule, one determines that the height of the hominid was 5 feet. The evidence of the fossils from Hadar and Laetolil shows 3 to 4 million years ago hominids were widespread, sexually dimorphic, and small-brained, with primitive teeth, jaws, and craniums. The Laetolil footprints show that the unique striding bipedal locomotion of people had been established before 4 million years ago (White 1980).

In Ethiopia, in a similar deep lakeside sediment in which erosion has carved countless gullies, thirty-four skeletons of *Homo erectus* have been found together, including that of a 4-year-old child. This site is probably a graveyard, for there is no evidence that they died simultaneously, but the experts have other theories. The adults were about 5 feet tall. R. L. Leakey and R. Lewin (1978) speculate that this band died suddenly and in a group; the hypothesis of drowning in a flash flood is ruled out by the geology of the deposit; perhaps there were poisonous gases from a volcano. This collection of skeletons provides the opportunity to compare relative dimensions of legs, arms, and spines with those of modern *Homo sapiens*. The hands were similar to those of today, judging from a complete hand pieced together from bones of several individuals. Their skeletons were also similar.

About 1.5 million years ago, hand axes and cleavers of weights up to a few kilograms appeared in use, knocked off as large pieces from even larger rock cores (obviously with the help of hammers long since in use). At the same time, tons of foreign stones were brought from elsewhere and accumulated at the tool factories as raw material. It seems clear that skin bags and vegetable fiber baskets must have been made and used for rock transport as well as for bringing in food.

The slaughtering and butchering industry concentrated its efforts mainly on bovine animals (43 percent), mostly dik-diks and other small antelopes, giraffes (11 percent), and rhinos (10 percent), the percentages having been deduced from some six thousand fossil skeletons in the bone piles (Leakey 1978). Besides these animal bones, there are also bones of hippopotami, elephants, horses, frogs, and fish. One can understand how a few humans could throw down and kill a giraffe or even a horse, cutting the jugular with stone blades, but how one subdues a hippo or an elephant and cuts through the blubber to the vital vessels requires a more complicated scenario. Probably these animals were driven to hidden pits, where they fell in and exposed the defenseless backs of their necks to the hunters' knives. Firebrands were probably used in driving the animals, although there is only inferential evidence of the use of fire in some pieces of welded tuff, showing what appears to be alteration by heat (Clark and Kurashina 1979). This dig, that of J. D. Clark and H. Kurashina, was made on the high plateau of Ethiopia at elevations of 2,300 to 2,400 meters, where the vegetation of the time, much like that of today, was grass with juniper, acacia, and podocarpus forests, as well as a river and a drying-up lake. The dates are 1.5 to 1 million years B.P., suggesting that hominids moved there as the Olduvai region began to dry up. If they used fire, one cannot assume that there should be residual charcoal, because it is not known if charcoal would survive for 1 million years (Libby 1979) from small deposits such as cooking hearths left by hominids, although charcoal does survive in bedded coal (Cope and Chaloner 1980).

Baked material provides evidence of fire at Chesowanja near Lake Baringo. Pieces of baked clay ranging in size from tiny flecks to lumps of several centimeters were found intermingled with artifacts and bones and only in such associations; these could not have been introduced into the site after its formation. Although natural phenomena such as lightning might have produced the deposits, Gowlett et al. (1981), studying the site and its surroundings, concluded that the remains strongly suggest that hominids of 1.4 million years ago were using and controlling fire.

At the Clark and Kurashina site, there are several fragments of heavily weathered basalt, which when rubbed give a red pigment. Thus there is inferential evidence that painting may have been a practiced art, but the evidence is inconclusive. The preferred meat seems to have been hippo; one of the floors bears part of a hippo butchery. Thus again, slaughter and butchery and toolmaking were standard industries, and one expects that dried meat formed automatically and perhaps was eaten.

Mary Leakey discovered in the Olduvai sequence the earliest construction yet known: a stone circle, which may have been the base of a tent, with the stones used to hold down the lower edges of the skin coverings. There are also pits at the Leakey site, the uses of which are not known (Leakey 1979).

Homes were generally located near lakes or along sandy beds of seasonal watercourses, where trees provided shade and perhaps refuge from nonclimbing predators and water was to be had, even in the dry season, when one obtained it by digging holes in the streambeds. All the lowland, Olduvai-related localities have a similar history, namely, evolution of climate from wet to less wet, from trees and tall grass to savanna with short grass and less shade. The hominids seemed to evolve through a succession of subspecies (Coppens 1978), perhaps along with the climate.

The evidence of the industrial complexes suggests the existence of families and bands of families, sharing food and dividing skills and labor, specializing in skills. For example, in toolmaking, evidence from neighboring living floors shows that tools at one site had characteristic differences in mode of flaking, suggesting that artisans were also artists, each with personal idiosyncrasies. Collaboration of the families and groups of families in these industries and skills suggests that verbal communication existed and some form of government operated (Harris and Herbich 1978).

There are other hominid sites as old as those at Olduvai, but only the Olduvai sites are firmly dated with K-Ar. Caves in South Africa at Sterkfontein evidence some tools similar to those at Olduvai, dated by cross-checking animal fossils there with similar animal remains at Olduvai. The oldest dates obtained in this way are 1.9 million years. This hominid, identified as *Australopithecus*, the same as at Olduvai, overlapped with and apparently evolved into a later species, *Homo erectus* (Howell 1965).

In a similar way, a site at Tell Ubeidiya, west of Kibbutz Reit Zera on the shore of Lake Tiberias in the Jordan Valley, has been assigned to the early Pleistocene, according to the context of fossil animals now extinct. Fragments of human skull found with them are younger, as measured by their fluorine content. Crude stone tools, choppers and cutting tools, were found together with bones which had been split to allow extraction of the marrow and which showed signs of scraping with flint knives. A hippopotamus jaw found with such tools indicates that hippos from Lake Tiberias were part of the diet and that group hunting must have been practiced to kill and use such big creatures (Stekelis et al. 1960). One may suppose that, as the climate of the Olduvai region and all of eastern Africa became more arid and the lakes and rivers dried up, the hominids of that region gradually moved to the north—where hippopotami, their favorite food, were abundant in the headwaters of the Nile—and down the Nile, from which it was an easy transition to spread into Israel, following the track of the hippo. Unfortunately, the Ubeidiya excavation was interrupted by terrorist raids in the sixties and seventies, and its continuation remains uncertain.

Over seventy sites in East Africa have now yielded remains of hominids of the Pliocene/Pleistocene (ca. 3.5 million years and after); the oldest yet found dates from about 3.68 million years in the Omo Basin, Ethiopia (Howell 1972). The dates obtained by all the various researchers have been reviewed by Hay (1980).

Cultural and Climatic Changes during the Pleistocene

G. L. Isaac (1972) has defined the Pleistocene as extending from 4 million years B.P. up to the Holocene, 10,000 years B.P. The splendid K-Ar documentation of dates of hominids of the Olduvai region of 1 to 4 million years B.P. hardly prepares us for the disappointing lack of dates for the time range 1 million years to 100,000 years B.P. It is only at 50,000 years B.P. and later that radiocarbon dates begin to define a time scale on which cultural changes may be charted and the rate of cultural change can be estimated. One reason for this lack appears to be that, although the sediments of Olduvai have been compacted into a remarkably tight formation, able to retain argon from potassium decay quantitatively or without significant loss, this desirable property is rare in most sediments.

After the deterioration of climate in East Africa, to the point that the essential diet of hippopotamus disappeared with the drying-up of the waterways, and the elephants disappeared with the change of abundant tall grass and trees into short grass and thorns, people also migrated elsewhere—apparently both to South Africa, where undated sites exist, and to the north, coming into southern Europe. Perhaps such sites as the yet undated Israeli Ubeidiya were left in their path, although how they circumnavigated the Mediterranean is not known.

Leakey and Lewin (1978) classify the hominids of the lower Pleistocene (2 million years to 0.5 million years B.P.) in several groups, all of whom walked upright. Those they believe to be the most primitive are designated as *Australopithecus robustus* and *boisei*, about 4 feet tall, with a cranial capacity of about 500 cubic centimeters (compared with the modern average of about 1,360 cubic centimeters). These hominids are dated at no less than 1.2 million years and perhaps more than 1.6 million years (Walker and Leakey 1978). As of 1978, sixty partial skeletons including several skulls had been found in former lake margins and former streambeds.

Australopithecus had a ridge of bone running back to front across the top of the skull, connected to a heavy bone ridge over the eyes; very wide cheekbones; and processes where the ears attached. The upper jaw projected markedly in front of the eyes and nose, and the incisors and canine teeth were large, although the front teeth were about the same size as those of modern humans.

The second most primitive has been named *Homo habilis*. This hominid had a cranial capacity of 775 cubic centimeters, lacked a skull crest, and had large front teeth, relatively smaller molars, and less prominent cheekbones.

The third, called *Homo erectus*, with a cranial capacity of about 800 cubic centimeters, had brow ridges less noticeably protruding above a small face, relatively large front teeth, and small molars like those of modern people. This hominid is similar to specimens found in northern China and Java but is about 1 million years older than they are. The similarity of the *Homo erectus* of Kenya to those who lived thousands of miles away suggests that this *Homo* was stable and viable for at least 1 million years. A. Walker and R. E. F. Leakey believe that this is the hominid who made the stone tools found in the East Turkana strata in northeastern Kenya. K. W. Butzer (1964), however, believes that the fossil evidence does not warrant a mass of generic names applied, one to each fossil, and that the designation *Homo erectus* is probably applicable to all the fossil hominids in this time range.

At the time that these hominids were alive, Lake Turkana was higher, with a shoreline 10 to 15 miles east of its present location, and the northeast portion of Kenya had a climate wetter than now. The region was one of lush rolling grasslands, forests, lakeside rushes, and moisture-loving vegetation.

The archaeological record was buried and thus preserved by lake beach silts and silts deposited by streams. The fossil record at Lake Turkana suggests that the principal erect hominid types described above coexisted for perhaps 1 million years. This supports Butzer's hypothesis that they were a single species, *Homo erectus*. Then about 1 million years ago *Australopithecus*, with the remarkable crested skull, gradually became extinct, the dates being based on K-Ar dating. The absolute size of their pelvises (five found up to 1970) and other skeletal parts shows that Australopithecines possessed a body the size of a modern pygmy (Simons 1970).

Somewhat more than 1 million years ago, bands of African-born *Homo erectus* journeyed into Asia and from there east to China and west into Europe. The *Homo erectus* of the Koobai Fora in Africa of about 1.5 million years ago looks very much like the *Homo erectus* who lived in China almost a million years later, who is called Peking man and who although formerly called *Pithecanthropus* is now classified in the species *Homo erectus*. Since he was first discovered in Java, further specimens have been found there, dated in the middle Pleistocene, together with stone and bone tools and traces of fire. Those living in China, *Homo erectus pekinensis*, had a pronounced meat diet, and the charring of the discarded bones indicates that they cooked their meat. In Choukoutien, China, there have been found the remains of more than forty Peking men, with tools of stone brought to the site and charred animal meat bones intimately associated with traces of fire. Largely two-thirds of the bones were venison, but there were also elephants, two kinds of rhinos, bison, water buffalo, horses, camels, wild boars, saber-toothed tigers, leopards, bears, and a hyena. Peking man probably had wooden spears with fire-hardened tips, used pits to catch the larger animals, and depended on teamwork based on articulate speech to catch and kill animals larger, faster, and stronger than he was (Clark 1969). Evidence of fire is seen in layers discolored by burning and melting, mixed with ash and charcoal and refuse, sometimes in caves.

Other middle Pleistocene deposits of *Homo erectus* have been found as far north as Heidelberg. The cranial capacity of the Java specimens was about 860 cubic centimeters, those in China 1,075 cubic centimeters. The lower part of the face projected forward. The heavy eyebrow ridges and the flattened forehead would be strikingly evident were we to meet this species alive.

The tool-making ability of *Homo erectus* has been estimated to have developed as follows. At first, from a pound of flint they could produce about 5 centimeters of cutting edge; in the middle of the Paleolithic about 100 centimeters could be produced; and in the late Paleolithic 300 to 1,200 centimeters were produced (Butzer 1964).

Recently, road cuttings in Togo on the west coast of Africa have uncovered hundreds of stone artifacts like those of the East African hominids but undated. In several areas there are narrow grooves on the rocks, produced by polishing stone axes, at one of which was found a well-made, weathered, bifacial stone hand ax. The largest number of grooves occurs on a big rock outcrop near Tchamba, where a rock gong was found near the grooves. The rock gong is a large slab of granite which vibrates when struck along the edge, an example of early human musical invention. Similar gongs, often with several slabs exfoliated from a large perched rock, have been found in several parts of Africa, notably in Nigeria and Uganda. Some gongs produce as many as a dozen distinct notes from the different slabs. The abraded nature of the gong edges where they have been struck and the chords they sound indicate their musical purpose (Posnansky 1980).

The Spread into the Climate of Europe

During interglacials, Europe was almost entirely covered by forests, from Spitsbergen in the far north to the oak forests on the shores of the Mediterranean, as is evidenced by analysis of the records of pollen stratigraphy. The cover by forests and the total lack of open grasslands meant that large grazing ungulates, such as buffalo and musk-ox, could not exist there because of the lack of their characteristic food, mainly grasses and herbs of the open range. However, in principle, mastodons and other browsing animals could find sufficient low-level branches to feed on. Thus forest-dwelling *Homo* may have had only marginal animal life on which to feed small bands of people and little or no chance to vary a meat diet with vegetable foods.

During the coolest part of the last glaciation, 20,000 to 15,000 years B.P., the floral picture was quite different. An ice sheet extended from the northernmost borders of the continent, down over the Alps. The southern part of Europe was covered with tundra and other cold-resistant vegetation, which could have supported reindeer and musk-ox. The entire North Sea Basin was dry, connecting Great Britain with the continent, and all was covered with a single deep ice sheet. During the various glaciations, the fauna consisted of such ungulates as mammoths, elk, woolly rhinos, reindeer, lemmings, Arctic foxes, and Arctic birds. A unique Arctic complex of mammalian fauna developed during the Pleistocene, where none had existed in preceding epochs. This was preceded, however, by a gradual evolution of a unique complex of mammalian fauna of the temperate zone, which lasted also during the subsequent glaciations, when it was presumably pinched to exist in temperate regions close to the equatorial climates. Typically, forest-dwelling animals of temperate regions were barred by tundra bordering ice caps to the north and by grass-covered lands bordering deserts to the south, which kinds of grazing lands were easily inhabited and crossed by horses, steppe rhinos, elephants, antelopes, wild boars, giraffes, and cats. Presumably, development of the Arctic and sub-Arctic large fauna made possible *Homo*'s penetration into and existence in near Arctic conditions.

In the late Pleistocene the fossilized traces of human artifacts, structures, and food refuse became wide and thinly spread as if social groups were nomadic, following the herds to the glacier edges in central Europe in the summer and to the Mediterranean in the winter. At the maximum Pleistocene ice age in Europe, the continental glacier came as far south as central Holland and northern France. The herds of ungulates migrating yearly between these latitudes furnished the best caloric diet for humans, providing the most meat in proportion to the effort and skill required to hunt them (Butzer 1964). This way of life is in contrast to that of a population dwelling on lakes and rivers, eating hippopotami which resided year round in the same place. In Europe, with its encroachments of mountain glaciers, the sediments containing late Pleistocene (less than 1 million years B.P.) dwelling sites and artifacts have been much too disturbed by seismicity and by glacial meltwater as well as by abundant rainfall to have produced residues compact enough to have retained argon, and so these are largely undated.

In Europe, there were herds of antelopes, wild horses, woolly mammoths, woolly rhinos, steppe bison, and giant elk. The woolly mammoth stood 3.5 meters tall, with tusks as long as 4 meters. It must have been difficult to kill this creature, requiring the hunter to cut through ⅔ meter of outside hair, 10 centimeters of undercoat, and 10 centimeters of fat to reach the jugular. By 35,000 B.C., the woolly mammoth had reached the peak of its population and was ranging from northern Europe across Eurasia through Alaska and into New England and even south to Florida, thus providing a food trail for the human invasion of North America from Asia during the late Pleistocene. The woolly rhinoceros, 3.5 meters long, standing 1.6 meters tall at the shoulder and with one or two dangerous meter-long nose horns, also ranged through the cold parts of Eurasia, as did the giant elk and bison, but not into North America, although the bison did.

Both the elephants and the rhinos became extinct during the last glaciation, but they lasted long enough to have their portraits painted in the caves of Lascaux, radiocarbon-dated at 15,000 years B.P. (Libby 1952), and at Altamira. Whether humans hunted them to extinction is a matter of continuing study (see Martin manuscript).

After the appearance of hominids in Olduvai, Pakistan, China, and Java, a million years ago and before, there are very few dates during the next 900,000 years. Seven samples of tuff from a lava flow, all dated by K-Ar at about 0.43 million years B.P., underlying an Aechulian stone tool industry near Rome (Howell 1966) are considered to be the most securely dated archaeological finds from the middle Pleistocene (see also Isaac 1972).

In 1980, a hominid molar tooth with early Neanderthal resemblance was found in a cave in Wales, thought to have been occupied between 200,000 and 175,000 years before the present. The date was determined both by uranium daughters and by thermoluminescence of a burned flint core found in the rubble surrounding the tooth. Other hominid remains consist of fragments of skulls found in Kent, dated at 200,000 years before the present. The tooth was embedded in stratified remains of such cold climate mammals as bear, beaver, vole, lemming, red deer, and reindeer along with hand axes, scrapers, flakes, and cores (Green et al. 1981). After that we have no dated sites until 100,000 years B.P.

The earliest Pleistocene occupation sites found anywhere (Klein

1974) in the European U.S.S.R. are roughly between 75,000 and 80,000 years old, dating back to the time of the last interglacial. It is possible that older sites will be discovered there some day. In Hungary, Poland, and Czechoslovakia traces of early humans have been discovered that are hundreds of thousands of years earlier.

So it is possible that only at the end of the last interglacial did humans learn to survive the harsh sub-Arctic climate of the Ukraine. The culture of 100,000 to 35,000 years ago is known as Mousterian. Its major accomplishment, the construction and use of heated houses, was more fully evident in the European U.S.S.R. than elsewhere, as found by the study of some one hundred Pleistocene sites in and around the Ukraine.

Many of these sites lie in valleys of the main rivers, because rivers accumulate sediments, and these bury and preserve the evidence. Many of these have been uncovered by road builders and diggers of construction material in sediments of sand and silt in the floodplain of the Dniester.

Pollen analysis shows that during these occupations the region was vegetated by steppe plants able to withstand cold and semidesert conditions, underlain by permafrost. The average January temperature was below freezing, but summers were warm although short and hay grew well. Winters were dry so little snow accumulated, allowing some thirteen species of herbivores, large and small, to graze and flourish. Favorites for hunters were reindeer, horse, bison, woolly rhinoceros, musk-ox, aurochs, saiga antelope, red deer, roe deer, giant deer, moose, wild goat, and sheep. Carnivores eaten were Arctic fox and wolf; no bear or lion was represented. Fish and birds are absent from the garbage heaps, perhaps because the occupation sites were used only in the winter when streams were frozen and many birds had gone south. Perhaps people summered on the uplands, following the grazing herds.

Mammoth bones mark the living sites in large numbers. Eight sites in the Desna Basin contained bones from five hundred individual mammoths. The evidence is that the bones at any site were gathered from long-dead skeletons as construction material. The most desirable bones for house supports were skulls, tusks, mandibles, pelvises, scapulae, and long bones. Wolves, foxes, and hares were skinned for their pelts, as evidenced by piles of their skeletons in one garbage heap and their paw bones in another heap. In the rings of mammoth bones that mark the ruins of the houses, there are quantities of charcoal and burned bones indicative of fires and hearths, for heating and cooking. Some of the ruins had floor spaces of up to 50 square meters, with many separate hearths and with hundreds of animal bones in the floor and thousands of pieces of flint, shaped as scrapers, saws, and points, and residues from their manufacture.

The houses may have been made of animal skins stretched over wood frames or over bones and tusks, with more bones and tusks holding the skins to the ground around the edges. No human remains have been found, but from the nature of the stone tools it is assumed that the *Homo* who constructed and used these sites was Neanderthal. There also are many artifacts made of bones and antlers, shaped as points, awls, and what may even have been needles, suggesting that these people made clothes, probably of skins. The suggestion of wood frames to support skin tents for the houses is borne out by recognizable holes to accommodate wooden uprights in the Dnieper-Desna Basin.

Whatever happened to Neanderthal is not clear. The archaeological record shows that at a certain level the Neanderthal tools were superseded by upper Paleolithic artifacts characteristic of Cro-Magnon. The latter are thought to have appeared first in southwestern Asia and southeastern Europe and to have spread out from there even to the Arctic Circle north of the farthest northern latitude (about 54°) at which Neanderthal sites have been found, apparently because they brought more successful adaptations to living in a cold climate.

Tool production developed in the late Pleistocene, in that artisans learned how to orient their core material so that they could rapidly produce dozens of extremely elongate, cutting-edged flakes, maximizing the length of cutting edge per unit weight of stone blade. Flakes with lengths greater than twice their breadth (knives) began to be ubiquitously produced in the late Pleistocene, about 100,000 years ago. Artisans had learned how to produce them in very large numbers by 40,000 years B.P., using two tools, a hammer and a punch, the hammer striking not the core itself but the punch, transmitting the force of the blow to the exact point where it could be most effective. Further invented was the process of finishing a tool by chipping it using pressure alone without percussion. Pressure flaking appears to have been independently invented in North America, producing the Folsom point. Better, lighter, and more easily obtainable sharp knives must have greatly increased the production of, and decreased the price of, slaughtered and butchered meat, bettering nutrition and increasing the number of people who could live together in such a way as to improve their standard of living.

Group size and organization are estimated from camp size, the number of individuals being roughly in proportion to the area. For example, an early Holocene hunting camp in Yorkshire, Star Carr, with a diameter of circa 15 meters, was estimated to have been occupied by twenty-five to thirty men, women, and children. Many sites coeval with the Olduvai are estimated to have been about the same size. Thus the evidence is that there was little change in group size since 1 million years B.P. until about 30,000 years B.P., when sites 50 meters in diameter occupied by one hundred to two hundred people began to proliferate.

Africa is littered with discarded stone tools belonging to the last 100,000 years or so. These tools show a marked reduction in size as well as a marked refinement for specific uses. Hafting of pointed stones into spears, javelins, and more easily used hand tools was invented at about 100,000 years B.P. and became an established part of the tool industry. Manufacture of the millions of stone tools dwindled only with the spread of the iron technology some 10,000 years ago (Isaac 1978).

The Neanderthal physical type, *Homo sapiens neanderthalensis*, evolved in 50 millennia or more and thereafter remained relatively unchanged for another 50 millennia, from about 150,000 years B.P. to about 40,000 years B.P. Then it changed into the essentially modern physical type of Cro-Magnon, namely, modern *Homo sapiens sapiens*. Or the modern human may have evolved apart from the majority of the Neanderthals. Or Neanderthal may be still with us, mixed among the Cro-Magnons and not noticed. The coexistence of two or more kinds of hominids has been found in the past as far back as 8 to 14 million years ago (Pilbeam 1982).

Physically, Neanderthals did not look much different from many physical types we see around us today. They were big and powerful. The brain was as big as or bigger than the brain of modern humans. The skull was lower and longer at the back on the average so that the forehead sloped on the average more than that of most but not all modern humans. The skull was distinctive, however, in that there was a gap between the wisdom teeth and the rising part of the jawbone—a distinction that would not have been visible from a cursory glance at a living person. There were a few other distinctive features, relating to the ankle joints, the bony structure where the pelvis fits to the sockets for the leg bones, and so forth, none of which would have been noticeable to an observer meeting a live Neanderthal (Trinkhaus and Howells 1979).

More than three hundred skeletons and partial skeletons of Neanderthals have been found in Europe, ranging from the Mediterranean shores to Britain and the Baltic Sea and east to Turkestan. But after 35,000 B.C. Cro-Magnons spread, absorbing and replacing Neanderthals across Europe and in the Near East. The people who went farther east and those who occupied Australia 32,000 to 40,000 years ago were modern *Homo sapiens sapiens*, even though this was at a time when Neanderthals still existed.

Neanderthals lived for the most part in open air sites, marked by masses of stone tools and rings of mammoth bones, probably used as poles for skin tents. Their remains are not as well preserved, therefore, as would have been the case had they lived in caves. And, thus, they also left no wall paintings in caves, nor do we know if they painted on skins, nor did they leave engraved bones or strings of shells and other kinds of beads. But they buried their dead and honored them with flowers.

Besides using axes and knives, Neanderthals hunted with wooden spears tipped with hafted stone points, as well as with stone balls probably used as bola stones, for entangling birds on the wing and for snaring running game. They also invented antler clubs and barbed bone points (probably for fishing with harpoons and trotlines), as testified to by fish and bird skeletons in their garbage heaps. Their arrowheads litter the arid parts of Spain and southern France.

Neanderthals achieved the first known human occupation of the sub-Arctic and Arctic, as well as of the African rain forest.

The Adaptability of Cro-Magnons to Many Climates

Cro-Magnons spread across Europe and Asia with remarkable speed. They occupied sites in Australia and New Guinea at 25,000 years B.P. and in New South Wales at 32,000 B.P., leaving cremated burial bones (Macintosh 1972). In reaching those islands, they demonstrated capability as the world's first known mariners; presumably they invented the industry of boat building at least 30,000 years ago. The English islands were, like Japan, at times connected by land to Europe. During the cold phases of the Pleistocene, when the great ice sheets accumulated on the continents and the sea level was lowered, the Japanese archipelago and the British archipelago were often connected to the mainlands. Japan was at times connected to Korea and at other times to Asia through the islands of Hokkaido and Sakhalin. Large Pleistocene animals such as elephants and megacerine deer crossed from the mainland, so it was possible for humans to do so; perhaps they were navigators as well. The earliest artifacts, hand axes, blades, and other Stone Age tools, have been found at dwelling sites radiocarbon-dated at 40,000 to 30,000 B.C., at Iwajuku about 90 kilometers northeast of Tokyo (Ikawa-Smith 1980).

In contrast, the sea separating Australia from Asia is too deep to have offered dry-land passage in the last 100,000 years. Therefore, passage had to be made by boat. In Australia, Neanderthals extracted flint from underground mines which they lit with torches. The flints served as raw material for the rapidly developing stone tool industry.

Cro-Magnons made remarkably rapid progress in the invention of tools and of furnishings for a more comfortable life. In the late Paleolithic of western Europe, about 25,000 to 17,000 years B.P., they produced bone and antler barbed points to be thrown on spears and harpoons and used as fishhooks. These were hardened in hot ash to make them durable and able to penetrate live-animal bone. Bone needles with eyes, and bone and ivory bodkins, were invented. The abundance of these artifacts suggests that clothing was well constructed, that skins and furs were routinely sewn into everyday clothes and probably into shoes, although the cave paintings do not show clothing and shoes on the hunters.

The bow with arrows, both untipped and flint-tipped with hafted arrowheads, was invented. Boomerangs and other throwing sticks as well as bolas came into daily use. Snares and pitfalls may have been known from earliest times, even at Olduvai. But in western Europe, as shown in cave drawings of southwestern France and in Spain, not only are snares, traps, pitfalls, and enclosures depicted but also masses of bones of herds driven over precipices and into pits have been found. Some of the cave drawings show, however, attacks on individual woolly rhinos, and other large animals, with showers of arrows (Butzer 1964).

Cro-Magnon people rapidly invented art and art appreciation. They made bracelets and beads and carved "Venus" figures and animals from ivory. They sculpted animals from chalk, modeled them in clay, and invented ceramic art by baking the clay. They invented engraving of bone, ivory, and amber, with geometric patterns and sometimes with anthropomorphic designs. They strung necklaces of deer teeth and amber and pebbles, using hardened bone awls to pierce the beads. For their own pleasure, Cro-Magnons painted on cave walls. This cave art has been found in Italy, France, Spain, and the Ural area (Clark 1969).

As far as the radiocarbon-dating evidence goes, Cro-Magnons were in the British Isles almost at the same time as in Europe. The oldest radiocarbon date for a British Cro-Magnon site is 47,000 B.P. ± 1,500 years for material found deep in a cave on the channel island of Jersey by Kenneth Oakley. The cave contained stone tools and the remains of cold climate fauna. A later Oakley find is a human skeleton in a cave on the Gower Peninsula together with the skull of a mammoth and stone tools, dated at 18,460 B.P. ± 340 years, about the time of the Weichselian glaciation.

In France radiocarbon dates show the Weichselian to have started at 70,000 years before the present, at which time deciduous forests disappeared. Sixteen radiocarbon dates have been measured for this lapse of time by which the continuous pollen record is calibrated; this now extends back 140,000 years in northeastern France (Woillard and Mook 1982).

However, radiocarbon dates for these prehistoric cultures are remarkably few in Britain. This is partly because meltwaters from the glaciations disturbed and destroyed the strata of cultural remains, partly because occupations must have been scattered between the tongues of ice and were successively iced over and revealed by melting, and partly because the longtime lapse in a wet climate has led to contamination of the organic materials which might otherwise have been carbon-datable. There are, however, four radiocarbon dates from material found in the Anston Stones cave and at Plixton, all of about 10,000 years B.P. with errors of about 100 to 200 years.

Along the Thames and in its adjacent valleys, there have been identified several sites dated at between 10,300 and 9,500 years ago. Many of these sites show land-to-sea relations different from those of today; these are supported in their radiocarbon dates by independent radiocarbon dating of their related pollen zones.

From the paucity of hominid remains in Britain, it is unlikely that the Paleolithic population there was ever very great even during interglacial times. People may even have spent summers in Britain and migrated to the continent for the winters (Molleson 1977), much as the jet set does today.

The establishment of Neolithic culture in large numbers of sites in Britain began at about 5,000 years B.P. or 3,000 years B.C., at about the same time as the Old Kingdom of Egypt was established. At this horizon there is the first appearance of pollen of agricultural weeds, which is interpreted as indicating deliberate deforestation to allow initiation of plant and animal husbandry in Britain. Then, by 4700 B.P., civilization suddenly became established. There appear a score or more of sites containing causewayed camps, henges, long barrows, pottery, flint mines to provide the raw material for the stone tool industry, and polished stone axes. Numerous wooden causeways were built beginning at about 4,000 years B.P., of which the pegs under the causeways are the most durable survivors of the structures. The wooden causeways undoubtedly served to keep people's feet dry and enabled travel over the bogs which had been initiated by glacial meltwaters and perpetuated by sinking of the southern part of Britain, caused by the rising of the northern part, relieved of the weight of glacial ice. The remains of the causeways are being discovered by peat cutters, whose progressing industry continues to uncover more examples (Allibone et al. 1970).

The British and Irish boardwalks found in peat mires, first dating from 4,900 to 3,900 years B.P., show the marks of stone axes only. Later walks, dated at 2,900 to 2,400 years B.P., show bronze ax cuts and square-cut mortise holes, a significant invention in woodworking, probably indicating the invention and use of bronze chisels. The bog stratigraphy bears evidence that walks were built during and in response to the climate worsening that occurred at this time throughout northwestern Europe at the so-called Sub-Boreal/Sub-Atlantic climate change when inundation may have been widespread, with colder climate and more rain. Dates for wooden trackways found in Irish bogs are 3395 B.C. ± 170 years; several are younger at about 1,900 years B.P.

Four English longbows have been found, an early one at 4625 B.P. ± 120 years of an unconventional design, namely, having wide flat staves bound with leather, and two later bows with British Bronze Age dates (Allibone et al. 1970).

The earliest prehistoric boats found in the British Isles are called "sewn" boats; they are elaborately constructed of heavy oak planks sewn together with stitches of yew and probably caulked with tree pitch and gum. The boats are thought to have come across the English Channel from Europe at 2,784, 2,796, and 2,305 years B.P., with counting errors of about 100 years. Younger ones are dugouts, two with steering boards. The boats attributed to European manufacture are comparable to boats found along the western coast of Europe, dated at about the same time (ibid.).

Pollen analysis, alike in England, Ireland, and Scotland, shows that the earliest forest clearings, of about 5,000 years B.P., were later abandoned and the forest regenerated in 200 to 400 years, to be subsequently cleared and reused two or more times for agricultural purposes. This was an early form of rotation of crops and may have been an inadvertent invention; it must have been unplanned, spanning as it did several generations of people (ibid.).

Archaeology during and since the termination of the last ice age has been complicated by the change in sea level. For example, a bone point was deposited on a North Sea beach in 8425 B.P. ± 170 years, then dry land, now under water. The rise in seawater took place between 14,000 and 6,000 years B.P.; the sea rose by more than 60 meters and by 5,000 years B.P. the rise of the ocean level from glacial meltwater was for the most part complete, both from continental ice cap melt and from melting of the ice cap on the British Isles. As the latter melted, northern Britain, relieved of the weight of ice, rose also; this to some extent negated the effect of sea-level rise in making beaches. The rise was uneven; down-warping caused production of the east fens (swamps and bogs) by about 4,500 years B.P. Objects found in association with buried trees are older than 4,500 years B.P., when peat growth in wetlands ended forest growth. This conclusion is verified by radiocarbon dates on pollen stratigraphy. This date also marks the onset of the Bronze Age (Godwin 1970).

At Ballynagilly there is a rectangular wooden house dated at 5168 B.P. ± 50 years, overlying a pit containing potsherds and pine charcoal dated at 5624 B.P. ± 50 years. The production of charcoal in connection with pottery sherds is significant for the discovery of metal production, as we shall describe. Furthermore, in Ireland there are passage graves as old as 4,600 years B.P. (Allibone et al. 1970). There is evidence for forest clearance in Ireland at about 5,300 years B.P. for agricultural and grazing purposes.

In Scotland the pollen index indicating the discovery and use of agriculture begins at 5,000 years B.P., with a counting error of about 100 years, and the dates for Scottish Neolithic sites confirm the age spread seen elsewhere for initiation of agriculture in the British Isles, namely, 6,050, 5,110, 5,030, 4,190, and 4,070 years B.P. and younger, with errors of about 100 years. These sites contain garbage heaps and such structures as chambered cairns, passage graves, and barrows (ibid.).

The successive stages of the Bronze Age in the British Isles comprise several sites dated at 3,800 to 2,800 years B.P. (ibid.), overlapping the invention of iron. The Iron Age is dated as beginning at 3,000 years B.P. in Scotland and in England and somewhat later in Ireland at 1,900 years B.P. It is hard to believe that the new iron technology took that long to cross the Irish Channel; probably older Iron Age sites will be found in Ireland.

The flint tool industry required raw materials, the supply of which developed flint mining into a major industry. In England and in Belgium, mines were continued underground as deep as 12 meters to get at the layers of flint embedded in chalk deposits. The mining tools used were picks made of deer antlers and horse scapulae. Flint mining became obsolete as copper and iron came into production.

The earliest occupation sites on the upper Nile date from about 15,000 years B.C., associated with stone flakes, burins (cutting tools with beveled edges used for engraving and scoring), and scrapers. The garbage heaps contain remains of large mammals and Nile fish, of which the latter are the more abundant. From this time on, the occupation sites were in use.

After about 15,000 B.C., the level of the Nile dropped, about 24 meters at Nubia but probably less in upper Egypt, leaving permanent ponds, silts, and deposits of diatomite. These indicate that the ponds were fed by seepage from the Nile and that the climate was cooler and perhaps rainfall was greater so that the ponds did not dry up as they would in the climate of today.

Similar to the erosive problems caused by the glaciers of Europe and the British Isles, one of the main difficulties in studying the earliest farmers of lower Egypt is that the Nile eroded or buried under silt most of the early occupied sites. The Nile flow increased and decreased over the millennia in response to variations of climate. Around the middle of the fifth millennium B.C., the climate was much more favorable than it is today. The Fayum Lake level stood 55 meters higher, and forest and swamp prevailed where desert exists today. In all the territory linked to the Nile Valley (a naturally irrigated valley), as far south as the Sudan and extending west along the southern margin of the Sahara, fish, crocodiles, and hippopotami flourished. In this environment the Old Kingdom and its high civilization developed. The Nile Valley, with its huge potential for growing food, gave birth to one of the great urban and literate civilizations of the ancient world, based on the development of a sophisticated art of farming (Clark 1969) providing plenty and leisure time.

In the first two centuries of the fourth millennium B.C. at Nagadeh in upper Egypt, there was a predynastic culture leading to the first dynasty, 3100 to 2890 B.C., which ruled over a unified country, effecting the unification of upper and lower Egypt under a single ruler. Before then, beginning in the Paleolithic, the earliest advent of humanity into Egypt was caused by climatic change across the whole of northern Africa. The grasslands and woodlands, lakes and rivers, slowly turned to desert, forcing the bands of humans to follow their means of subsistence, the herds of animals, moving ever closer to the great River Nile. Arrowheads found in the desert trace their movements, and flint sickles suggest that these people were learning that wild grains could be grown in the sediments of the Nile. Somewhat before 4000 B.C., people settled in the northern limits of the Fayum Oasis at Merimdeh Beni Salahem, and at Deir Tasa and Badari in upper Egypt, and at El Omari in lower Egypt, no longer nomadic and no longer dependent on the movements of wild animals.

In the predynastic times there were significant differences in the tool kits and household furnishings between upper and lower Egypt. In upper Egypt, flint arrowheads and granite hammers were manufactured. Clay pots were black or red, decorated with white incisions, sometimes with wickerwork, sometimes painted with animals and hunters. In lower Egypt, people were buried in tombs ceilinged with beams and carpeted with woven mats. The dead were supplied with pots painted with scenes of people and many-oared boats and with bronze statuettes and glazed stone beads.

Inhabitants of these predynastic settlements cultivated barley and flax, wove baskets from reeds, and made pottery of imaginatively varied shapes with polished surfaces and colored clays. The earliest Egyptian terra-cotta human figures, some with beards, date from this time.

At about 4000 B.C., a group of people immigrated into Egypt from an unknown place in Asia, coming across the Red Sea and by the Wadi Hammamat. It is thought that they brought the ideas and impetus for the foundation of the Egyptian Kingdom, unifying Egypt. The unification of Egypt, circa 3100 B.C., was accompanied by a supreme flowering of technology and art, supported by an abundance of food grown in the Nile sediments with a minimum of effort. With these phenomena—plentiful food, bodily well-being, a salubrious climate, and leisure—began the building of the great pyramids, the Sphinx, and the temples, in the periods of seasonal unemployment caused by the Nile floods (Michalowski 1973).

In southern Europe and in Asia Minor, where the glaciers did not penetrate, the record has been found from much earlier times. By 40,000 B.C., an essentially modern fauna had appeared in Asia Minor. By 35,000 B.C., the production of flint tools had become established over southwestern Asia and northeastern Africa, with some degree of bone working and stone polishing. By 10,000 B.C., there had appeared a variety of new tools of ground and polished stone, large grinding stones and mortars, more bone tools, and bone handles for flint blades, also rough attempts to model humans and animals in clay. The grinding stones indicate that at least wild grains were used as food, and perhaps the cultivation of grain had begun. This became certain by 7500 B.C., when village farming communities had been developed in southwestern Asia; effective food production was underway to enable permanent villages to exist (Braidwood 1970; Braidwood, Campbell, and Schirmer 1981).

The cereals cultivated in Syria included barley, wheat, and einkorn in 8000 B.C. The garbage dumps contained only about 5 percent animal bones, indicating a predominantly vegetable diet. There were few indications of animal husbandry in the Near East, but in England, at Star Carr, a living site dated at 9388 B.P. ± 350 years (7300 B.C.), were found the remains of a dog.

By 6000 B.C., stable villages could be found throughout Europe and far into Asia. The good conditions of preservation at sites in dry climates allowed the survival of wooden vessels, baskets, monochrome and painted pots, spoons, ladles, needles with eyes, handles of bone, mirrors of polished obsidian, animal skins prepared for clothing, and woolen textiles (indicating domestication of sheep or harvest of sheep wool from bushes and thorn trees), but still the tool technology was Neolithic, made of flint and obsidian.

After the domestication of plants, animal husbandry became a certainty (Protsch and Berger 1973b). Sheep, goats, and dogs were domesticated, and meat was added to the diet of grains; woolen textiles replaced skins for clothing; houses were built of sun-dried bricks (in climates that allowed mud to dry and bricks to be preserved) glued together with mortar containing ash and bone.

Living was also easy in the region which is now the Iraqi-Iranian frontier. At 7000 B.C., the village of Jarmo had multiroomed houses built of sun-dried bricks, with clay ovens and clay structures for storing grain. The garbage heaps and the artifacts remaining today show that the villagers depended only about 5 percent on hunting and subsisted on farming barley and peas and herding (domesticating) sheep and goats. Since Jarmo was close by marshes, its diet was supplemented by carp, turtles, mussels, and water birds (Clark 1969).

In the interval 15,000 to 12,000 B.C., grain, probably wild, began to be ground, as indicated by grinding stones. From 7000 to 3000 B.C., bone harpoons were used, grinding stones became ubiquitous, and pottery appeared (probably associated with baking bread, in the process of which dried clay pots were found to become hardened).

The pottery is "frequently tempered with fiber" and therefore not all of it was fired after sun-drying (Wendorf, Schild, and Said 1970).

The earliest metals were native gold and copper found by panning alluviums, using wooden pans and shovels much like those used today in streams of the western United States. Copper was already in use by Sumerians in southern Mesopotamia by 3500 B.C. and was common by 3000 B.C., by which time reduction of its ores had been achieved, and the art was rapidly transmitted to Europe. Peak copper production occurred in Asia Minor in 2400 to 2000 B.C., after which iron production became dominant (Wells et al. 1981). Goldworking, of course, was well known even in the Old Kingdom of Egypt, and copper-working arts spread into Egypt by 2600 B.C. Copper ore was mined both in eastern Egypt and in the Sinai, but later the art of copper production became more skilled and less expensive in Cyprus and Armenia; thus these places exchanged copper with Egypt, presumably for grain and perhaps cotton.

The Book of Job, dated tentatively at 400 B.C., describes the technology of reducing copper ore underground by setting wood on fire in the mines, producing a reducing atmosphere in somewhat the same way it is proposed today to burn coal underground in the mines of Sinai to form synthetic gas, methane and hydrogen, by the addition of water to the burning coal. In Europe, Bronze Age copper mines were developed in Austria, Germany, France, Spain, Portugal, Greece, southern Russia, and Switzerland. In Tyrol, underground burning was also used. In Bulgaria a vein of $Cu(CO_3)_2$ was completely worked with underground fire. The modern authorities (Bromehead 1954) ascribe to underground burning the purpose of splitting rock, including pouring water on the hot rock to allow better access to ore, but at the same time a reducing atmosphere was produced which reacted with the ore to produce metal.

Persia may have been the site of origin of the oldest metallurgy, from which it spread to the Near East, where a multiplicity of centers arose for the development of metal technology at places where ore and fuel (wood) were available. This view of the origin is based on Hittite historical documents. But iron was known and used throughout the eastern Mediterranean lands as early as the third millennium B.C. and became common in the mid second millennium. Tools, weapons, and jewelry were made of iron in forms which copied preexisting bronze examples, in many cases specifically local examples, suggesting that iron was worked locally rather than imported from a single center. Other examples of the widespread iron technology come from Ban Ching in northeast Thailand, dated before 1600 B.C., and from graves of the Shang period at K'ao Chang, China, dated around 1200 B.C. (Champion 1980). None of the ancient empires of the Near East, set in climates changing toward desert, was rich in timber, so trade developed in importing crude metals from distant smelters where timber existed, to be finished by local metalworkers.

It is thought that the first observation of reduction of copper ore (copper carbonates and sulfides) was made while baking copper compounds used as glazes in pottery kilns. Copper ore is reduced at temperatures as low as 800° C, and the resulting copper metal sponge is melted at 1,083° C, temperatures attained in early pottery kilns. The history of forced draft to obtain higher temperatures is obscure. But blowing on coals is natural to humans in order to start fires from tinder and to warm themselves. Reeds used as blowpipes to concentrate the air were an obvious accessory. Blowpipes were used in Egypt as early as 2500 B.C. Bellows of animal skins are shown in Egyptian paintings of 1500 B.C.

Both gold and copper occur in nature as metals. It is not certain that the use of gold preceded that of native copper: in Egypt native copper may have been worked earlier than gold (Forbes 1954). But extraction of ores of both kinds of metal developed into such a big business in Egypt as to become established as an economic trade monopoly. Gold was present as metal fragments in quartz veins and alluviums. These rocks, when crushed to powder, were washed away, leaving metallic gold. The yield of gold in Nubia is estimated to have been 30 kilograms per year, and of copper in all Egypt 20,000 tons were produced between 1300 and 800 B.C. Casting of copper and gold was discovered in about 3500 B.C.; bronze became common about 2000 B.C., the tin-bronzes having superior strength and hardness. Thus trade in tin ores developed into an important industry.

The true Iron Age began about 1200 B.C. in the Near East. The new agricultural tools made possible clearing of forests, drainage of marshes, and improved cultivation, with the result that grain became less expensive. Iron came into use much later than copper, despite the fact that their smelting temperatures are not very different. But, unless iron ore is heated to the melting point (fig. 3-2) while it is being reduced to iron, the metallic iron does not liquefy and run out but remains as metallic particles mixed with stony material from which it is separated only by hammering. However, in the period 1900 to 1400 B.C., heating, hammering, and quenching became recognized iron-fabrication technologies which, when metallic iron particles had been hammered into an ingot and the ingot was many times reheated in contact with charcoal, produced a high-carbon hard steel, suitable for weapons and heavy-duty tools for clearing forests, for agriculture, and for building boats, houses, and boardwalks.

One further invention involved the addition of suitable fluxes which formed solid solutions with the slag, lowering its melting point and enabling it to separate from the metal sponge at lowered temperatures (Forbes 1954). Recognition of the relationships between meteoric iron, high-carbon (up to 1.5 percent) iron fluxes, and soft wrought iron initiated the true Iron Age and the onset of the explosion of modern technology (ibid.).

Circa 800 B.C., use of iron was brought into the Ganza Valley via trade routes. It had been coming in bits since about 1000 B.C., in the form of arrowheads, daggers, spearheads, nails, knives, etc., together with red- and black-painted pottery (which has been found in sites dated at 4000 B.C. at the earliest). Iron as an item of trade is thought to have diffused from Syria after the conquest of the Hittites circa 1800 B.C. The Hittites learned smelting of iron, classified the technology, and made its manufacture and role a monopoly and a commercial success. After their defeat, their tribe, with its superior iron

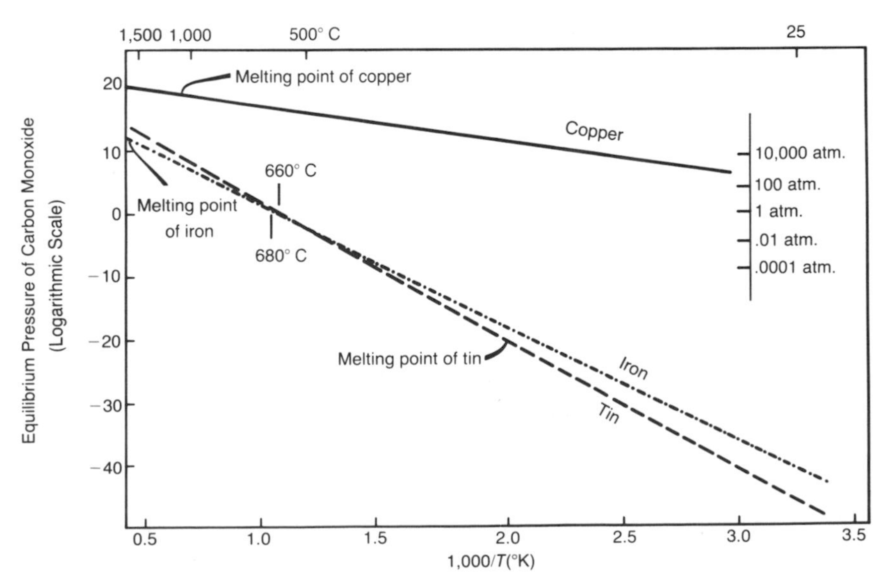

FIG. 3-2. Production of metallic copper, tin, and iron by the reaction of ores (metal oxides) with charcoal versus a function of the temperature of the oven used for the reaction and of the equilibrium pressure of carbon monoxide gas in the oven.

technology, was dispersed over Iran to the borders of India and into the Ganza Valley as refugees.

India had abundant high-quality iron ores and its artisans were adept at copper-bronze technology, so the prehistoric Indians could have developed iron earlier (Bromehead 1954). The temperature of melting copper (1,083° C) is too low to melt iron (1,535° C). At the lower temperature, iron was produced as metal bits in slag blocks, called blooms, which were hammered while hot to make the bits of iron stick together and to knock off the rocklike slag, whereas copper and bronze could be melted in the usual furnaces, cast in molds, and forged cold if need be. Forging of iron was enabled by blowing air on the fire with bellows to produce enough heat to warm and rewarm the iron blooms and forgings. The iron so obtained contains up to 4 percent silica, 2 percent alumina, 1.2 percent lime, and 0.8 percent MgO, with smaller amounts of phosphorus and sulfur. Of six iron objects, carbon content was less than 0.5 percent in all but one, an iron nail. These objects, dated at 600 to 400 B.C., could all have been made by forging the sponge of metallic iron produced by direct reduction of iron ores (oxides of iron) with charcoal. However, the reduction occurs only in the presence of an excess of charcoal; otherwise, if CO_2 was produced in sufficient amounts (owing to insufficient carbon present), the metallic iron was reoxidized, the experiment was identified as a failure, and the contents of the crucible were tossed out on the slag pile (fig. 3-2).

Iron ore can be smelted at 800° C and above to produce metal particles, but they are very porous, with the pores filled with stony material that comes with the ore, and they cannot be made to stick together by hammering. Iron smelted at 1,000 to 1,050° C can be forged, but with difficulty. Only at 1,100 to 1,150° C and higher does iron soften sufficiently to form a porous mass that can be forged, namely, from which the stony material can be ejected by hammering. At higher temperatures the stony material melts and drains away by itself, especially if fluxes are added to lower its melting point. Hence, forged iron objects produced at the dawn of the Iron Age imply charcoal-burning furnaces which could reach 1,100° C.

Meteoritic iron was known and was valued for jewelry but was of little use for making iron and steel tools because of its high nickel content, which rendered it nonmanageable at 1,100° C. All iron meteorites have nickel contents of 6 percent and greater (see, e.g., the appendixes in Wasson 1974).

Bronze was ever in short supply, because its components—copper mixed with a small amount of tin or mixed with small amounts of other alloying components such as arsenic and antimony—are in relatively short supply among natural world resources. Yet the evolving civilization of humankind had quickly adapted to metal agricultural tools, for more widespread and effective scratching of soil to produce grain more abundantly than ever before, and had quickly made known the pressing demand for more and less expensive metal tools. To fill this need came the appearance of the iron technology following rapidly on a series of inventions, probably fortuitous, certainly interdependent, and all based on the marvelous chemical properties of the ubiquitous element carbon.

That the iron technology could have been so rapidly developed, or even developed at all, probably followed from the new freedom derived from more abundant grain, now cultivated and no longer harvested from the wild. The abundance and cheapness of this food derived from metal agricultural tools and, no less, from the spectacularly kind climate which developed as the glaciers retreated, starting about 15,000 years B.P., and freed a small fraction of the population from the daily hunt for food to devote some time to invention. Invention of new technologies has always depended on a certain amount of leisure, deriving from a relative abundance of the necessities of life—water, food, warmth, and safety.

The inventions of the iron technology followed from the smelting of copper ores to make copper. The oldest copper smelter sites known are in Iran and Israel, dating from the fifth and fourth millennia B.C. The commonest ore, copper iron sulfide, must first be roasted in air to volatilize sulfur as gaseous sulfur dioxide, leaving the ore as mixed solid oxides of copper and iron, after which the oxides (ores) are smelted. This is done by filling a stone furnace with alternate layers of charcoal and ore mixed with a flux. A widely used flux, even today, is sand, which combines with metal impurities to produce lightweight silicates which float on top of the melt and hence are easily separated by skimming the melt or banged off after the metal has solidified.

The charcoal layers in the furnace were ignited and the heat was increased by a natural draft of air flowing in from holes at the furnace base, dragged through the furnace and out the flue at the top by the evolving hot gases. By regulating the amount of air coming in, the smith could restrict the evolved gases largely to carbon monoxide. The reaction of the oxides in the ores with carbon (charcoal) and carbon monoxide thus reduced copper oxide and iron oxide and other metal oxides present in the ores at temperatures available in the furnaces of those times, about 1,100° C. The abundant molten copper trickled to the bottom of the furnace, leaving most of its impurities behind as silicates in the residue, called gangue. Of course some copper was lost to the gangue also, and some impurities, reduced to metal, failed to be removed as silicates and trickled along in small amounts with the molten copper, producing natural bronzes.

In banging off the frozen gangue from the solidified bronze, the smith soon found that the bronze edges were hardened and toughened. Thus each great bronzesmith held to his own special formula and sequence of operations—the mix in the furnace, the rate of admission of air, the height of the flue, the shapes of the molds in which the trickle of metal was cast, and the amount of hammering on the cooled metal shapes. All these were trade secrets.

But copper, and therefore its bronze alloys, became scarce; perhaps the mines from which ore could be transported by donkeys and camels became exhausted. So the production of iron metal from iron ore became a pressing need, iron ore being abundant almost everywhere. But pure iron melts only at 1,535° C, far above the 1,200° C or so obtainable with charcoal and air in the beehive furnaces of the day. Smelting iron at that temperature, with layers of charcoal and ore and sand through which air flows or is forced, yields an iron metal sponge mixed with gangue: iron oxide and iron silicate. The smith (probably instinctively) hammered the sponge to squeeze out the gangue in between successive reheatings of the sponge on charcoal. If there were few exposures to surface absorption of charcoal, the smith produced soft iron; if the iron were reheated on charcoal many times and absorbed as much as 1 percent carbon, the surface of the iron tool being hammered turned to steel and became very hard.

The soft iron so produced is a soft impure metal streaked with grains of rock, of a tensile strength greater than that of copper but only slightly greater than that of bronze, which, however, could not be cast in molds as copper and bronze could because the smith's furnace was not hot enough to make liquid iron. But the remedy of this problem was probably discovered serendipitously, namely, the soft iron was made into steel. Every time the smith reheated the sponge for further hammering, he rested it on white-hot glowing charcoal in the forge, surrounded by an atmosphere of evolving carbon monoxide, with the result that the soft iron mass remained reduced metal and gradually dissolved carbon in its surface, converting the surface into carbon steel, which when cold-hammered could be brought to twice the tensile strength of bronze. Quenching after hot-hammering was likewise discovered to increase the hardness of the metal by preventing the growth of large metal crystals.

After 900 B.C., the production of steeled iron tools blossomed; thousands of these implements have been found in northwestern Iran, in Iraq, and at many other sites in the Near East. It was much later, in about 500 B.C., that the Chinese discovered that, by dissolving as much as 4 percent carbon in iron, the melting point could be reduced to about 1,150° C and the resultant steel could be cast—but at a cost of producing a very brittle material, so that this discovery, although important to industry ever since in the production of castings, was of no use to the techniques of forging steel (Madden, Muhly, and Wheeler 1977).

Human Mitigation of Climate: Shoes

In less than 50,000 years, humans extended their range of living sites into Europe and Asia, even crossing wide, deep waters to populate Australia and the string of islands in between it and the Asian mainland. They invaded the cold regions of the sub-Arctic, and they made their homes and provisioned them by hunting up to the edges of the ice of the European glacial times. One suspects that they must have invented and developed shoes to assist in their swift geographical conquests. But we have as yet no firm evidence, no footprints since the prints of bare feet were made and preserved at Olduvai some 3 million years ago, and no cave paintings of shod hunters.

The major human invasion of the Americas (fig. 3-3) took place rapidly; human migrations swept from the Aleutians in the far north to the southernmost tip of South America in about 1,000 years. Shoes would have been of enormous comfort as people crossed from Asia to Alaska, walking in ice-cold water and on icy tundra, and would have greatly assisted their progress south over the lava flows, deserts, hot beaches, and rocky shores which characterize the Pacific coast.

Now, for the first time, there is evidence of ancient shoes, not of just one shoe but of a whole shoe industry. A cave containing seventy-five sandals, including nine children's sandals, was found in the excavations of the caves of the Fort Rock Lake area of south central Oregon (Cressman 1942, 1966, 1973, 1977; Bedwell 1970, 1973). Sandals have been discovered in all the caves excavated, the most in Fort Rock Cave itself (figs. 3-4 to 3-6), dated at 9053 B.P. ± 350 years (Libby 1955), and several more in Catlow Cave. Three types of these woven sandals come from the south central Oregon caves; from Fort Rock come two types, the multiple warp and the spiral weft. It is

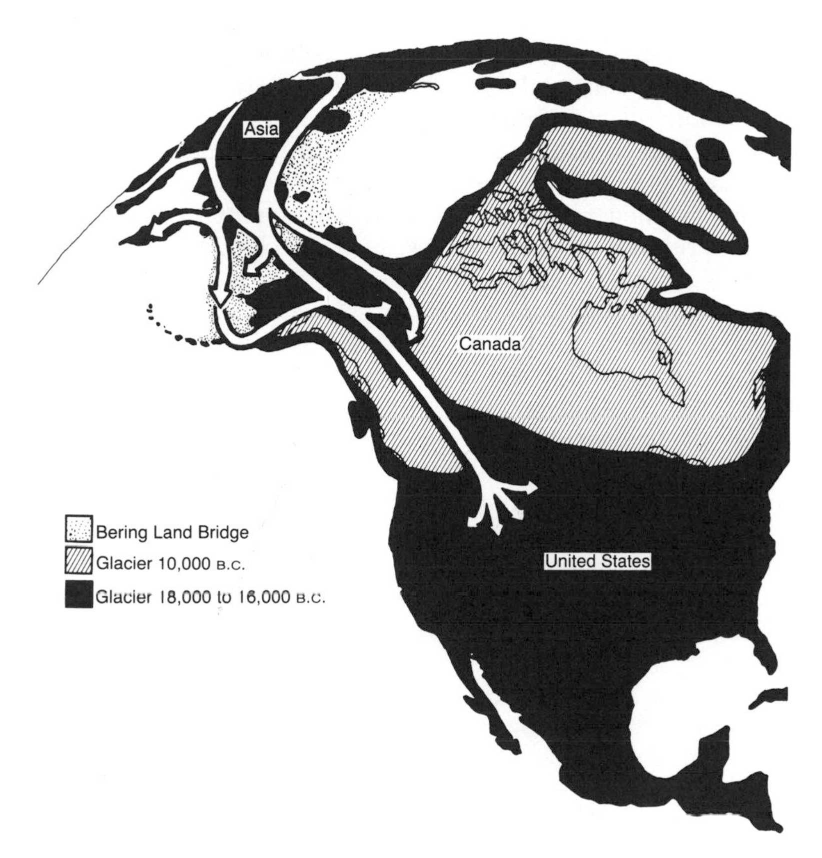

FIG. 3-3. Hypothetical early human migration routes from Asia to North America, made possible by a 150-foot lowering of the sea level caused by the buildup of 1,000-foot-deep ice sheets on the continents.

surprising that there were only three types. Some are so beautifully woven that they would be in demand today for wear around the house and garden. They are equally designed for left or right feet, so that, if one shoe wore out on the glass-sharp lava flows of the surrounding countryside, it could be thrown away and a new one substituted. Apparently hunters carried sacks of shoes with them. All the Fort Rock sandals were found in a layer of pumice and ash below the last lava flow of Mount Mazama, dated at 7000 B.P. (ibid.), and are somewhat charred from the heat of the pumice and ash.

The eruption of Mount Mazama, although about 55 miles southwest of Fort Rock, nevertheless deposited about 6 inches of hot pumice on the area and in the caves themselves. The effect on people and plants and animals was therefore potentially great. As well as pumice fall, such eruptions cause widespread volcanic fumes of hydrogen sulfide, hydrofluoric acid, and sulfur dioxide, which react with atmospheric water vapor to produce acid rains. We have destructions by modern-day eruptions as examples of how Mount Mazama's eruption may have affected Fort Rock. After the eruption of Mount Katmai in 1912, an acid rain fell at Seward, 250 miles from Mount Katmai, and at Cordova, 360 miles distant, painfully burning people and vegetation, charring wood, and corroding metal. Fumes near Cape Spencer, 700 miles distant, were strong enough to tarnish ships' brass in 20 minutes.

Many of the charred forests and burned vegetation of past volcanic eruptions, attributed to hot ash falls, may have been produced by low-temperature acid fumes and acid rains. The Mazama eruption, estimated to have been much greater in energy than that of Katmai, based on the amounts of material expelled by the two volcanoes, would certainly have caused similar destruction over a wide area. The same disaster must have followed the volcanic eruptions at Olduvai.

The 8-inch ash fall itself would have smothered low-growing plants and killed larger ones by mechanical overloading and chemical corrosion. On Kodiak, after the Katmai explosion, a 10-inch ash fall destroyed all low-growing vegetation and produced soil toxic to germinating seedlings for many years after the event. Lakes and marshes, important to the Fort Rock inhabitants for fish, water, and tule reeds, from which they wove baskets, would have become clogged by ash and acidified by rain, smothering and poisoning the biota. The rivers and streams would have been slurried with ash year after year as the pumice was gradually washed from the uplands. With food supplies wiped out, the Indians probably had to move elsewhere.

Animals also are vulnerable to hazards of volcanoes. In the case of the Hekla eruption in Iceland, animals grazing on the ash-covered pastures suffered tooth and joint damage as a result of fluorine poisoning. Cattle on Kodiak, after Katmai, were removed to the United States and brought back two years later after the pasture had revived. After the Parícutin eruptions in Mexico, ash rapidly reduced the local animal population. The smaller animals were smothered or starved and deer, rabbits, and coyotes disappeared as their food supplies were destroyed.

In April 1980, Mount Saint Helens, in the same chain of volcanoes as Mount Mazama, shot steam and ash plumes 20,000 feet into the air. Ash fell as far north as Vancouver, Washington, as far south as 50 miles to Portland, Oregon, and east to Moscow, Idaho, 300 miles away (*Los Angeles Herald Examiner*, April 1980), thus displaying a modern example of what the people of an older time in the neighborhood saw and felt.

On May 18, 1980, Mount Saint Helens exploded with a thud heard 100 miles away, accompanied by an earthquake of Richter 6.5 at a depth 10 miles below sea level. The explosion blew 1,500 feet of rock off the top of the mountain. The signature of the quake on seismographs differed from usual quake signatures in that there was a much larger proportion of long-period vibrations able to travel around the world three times before damping out. Five dead were found in two cars that had been overrun by pyroclastic flows, rapidly moving flows of hot rock ash riding on superheated gases that may be as hot as 700° C. The flows mixed with snow and ice to form giant mud slides that knocked out steel and stone bridges and buildings and engulfed other buildings in up to 30 feet of mud. The explosion left a 1-by-2-mile crater. Trees were snapped off at ground level up to 6 miles away, totally clearing a timber-covered area of forest of about 120 square miles. Four million board feet of logs were caught in the fast-moving torrents of mud and ice, filling a 20-mile stretch of the Toutle River. Water in the river and smaller streams on the flanks of the mountain reached temperatures as high as 90° C, cooking the fish. Fish that weren't cooked were smothered with floating ash and engulfed in mud. Millions of salmon were destroyed and their spawning grounds devastated. Dead deer and elk were seen on the mountainside by reconnaissance aircraft; farther away dazed animals stumbled about in the overturned forests.

Water service officials warned against drinking water that came from exposed reservoirs, which were covered with rock dust and wood ash from the many forest fires ignited by hot volcanic ash. The Columbia River between Oregon and Washington was closed to ship traffic by a 25-foot underwater bar of mud and by a 20-mile logjam of trees and smaller debris. Streetlights turned on automatically as darkness overtook daylight in the middle of the day. Hunters and fishermen abandoned their equipment and canoes in order to escape as rapidly as possible to save their lives. Snowplows were brought out of storage to clear the hundreds of miles of ash-covered roads. The railroads, Burlington Northern, Union Pacific, and Amtrak, halted some rail operations in Washington State, Montana, and North Dakota; western airlines curtailed flights in Wyoming, Canada, Montana, and Washington. Federal air authorities closed off 1,800 miles of airspace from commercial airline traffic around the mountain.

Agriculture authorities speculated that the wheat crop would not be hurt if the ash fall didn't exceed 2 inches, but the 4-inch and rising ash fall in the Yakima Valley threatened cherries, apples, and peaches. Farmers were sweeping ash off livestock and plants because attempts to wash it off only produced a thick sticky mud.

Easterly winds carried ash over western Washington State, Idaho, and Montana, blanketing the area with ash falls that resulted in the closing of entire communities and highways. Automobiles were unable to operate because their filtration systems clogged and caused overheating of the cooling water. Those that managed to operate for a short time raised such clouds of ash as to reduce visibility to only a few feet. The Spokane airport was closed to allow for scraping the runways, because the ash made them slippery and rose in clouds that affected visibility. The National Oceanic and Atmospheric Administration predicted that global weather patterns could be altered for months to come and that suspended particles might stay in the stratosphere for 2 years, lowering temperatures in the northern hemisphere.

Thus, in the Fort Rock area, the entire life chain was probably exterminated by the volcanic catastrophe. There was an abrupt cessation of grass production as determined from the pollen record, which is, however, not precisely dated as are the layers under the pumice in the caves; it may well have been caused by the Mazama ash fall. Even a change in climate could hardly have so drastically and suddenly decreased the growth of grass (Bedwell 1973). But in between explosions people lived a good life and developed an impressive civilization.

The Fort Rock sandals (figs. 3-4 to 3-6) are woven of shredded sagebrush: bark shreds twisted counterclockwise into a rope which runs first racetrackwise the length of the sole into a series of "warps" and then crosswise into a series of wefts woven above and below the warps. The length of the sole is made about 20 centimeters too long for the human foot and is turned back and lashed to the sole on the edges to form the top of the shoe, with a decorative braid to go over the arch of the foot. There is no heel.

At that time, although the climate was desertlike, there were abundant lakes in the region which had been formed by meltwater from the great ice sheets covering western Canada and Washington State; these many bodies of water supported a large population of the

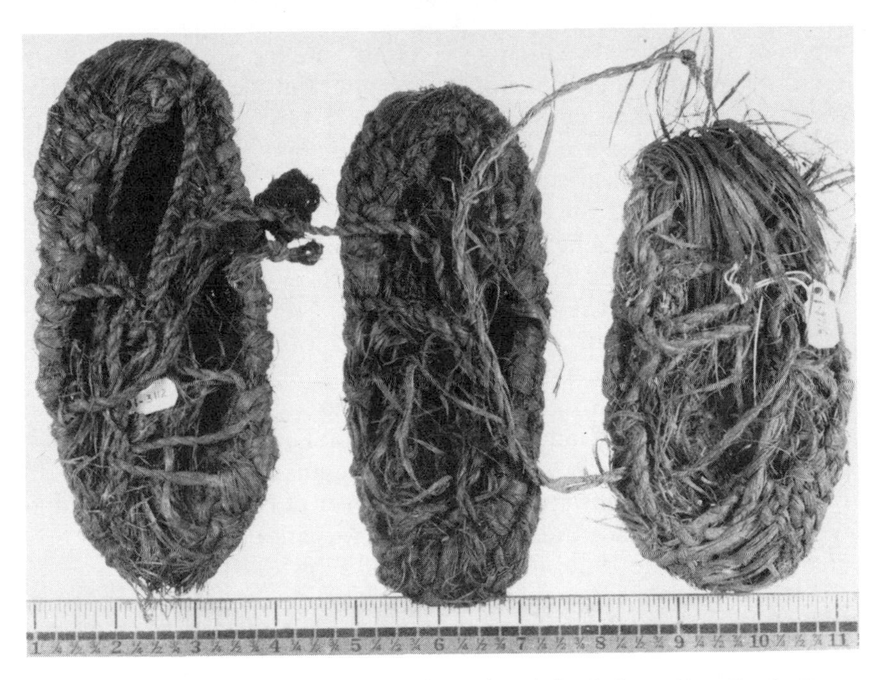

FIGS. 3-4 to 3-6. Sandals woven of sagebrush bark from Fort Rock Cave, Oregon, radiocarbon-dated at 9053 B.P. ± 350 years (Libby 1955).

water birds which formed part of the diet of the Fort Rock Cave people and provided forage along their edges for the large mammals which were also part of the diet. Drying-up of the lakes in the next thousand years caused a profound change in the plant and animal population and in the amelioration of climate caused by abundant water reservoirs and thus affected the number of humans able to live off the region and their way of life, their diet, and their health.

Evidence for rapid climate change occurring at that time somewhat to the south, namely in the Mohave Desert, is provided, for example, by researches of Rainer Berger and P. V. Wells (1967). They analyzed the pollen content of sloth (now extinct) dung found in caves in the Mohave Desert, radiocarbon-dated at 10,500 B.P. ± 350 years, and found significantly more pine, juniper, and artemisia pollens, suggesting cooler and moister conditions than exist today. Pieces of plants found in the dung were identified as leaves of the Joshua tree (*Yucca brevifolia*), which no longer grows near the caves but does occur at higher and cooler elevations about 70 kilometers to the southeast. Their study of pack rat nests dating from 10,000 years ago identified remains of Utah juniper (*Juniperus osteosperma*) and pinyon pine (*Pinus monophylla*), demonstrating that then the higher areas of the Mohave Desert, now occupied by scanty desert shrubs, supported woodlands of pygmy conifers, indicating cooler temperatures and much more rain.

Similarly, along the edges of Hudson Bay, Rhodes Fairbridge has found a series of beaches caused by the sea level falling during glacial stages and rising during interglacials, for some seven glacials and interglacials in the last 400,000 years. Profound changes in climate accompanied this succession of glaciations, affecting the temperate and tropical regions as well as those nearer the edges of the advancing and retreating ice caps. The migration paths and summer grazing lands, nesting lands, and procreation lands changed with each glaciation, for birds and the large grazing animals on which small bands of humans subsisted; the plant cover of necessity changed as well, and the surface area of dry land was in a continual state of change (Fairbridge 1960, 1961; Etkins and Epstein 1982).

These people were appreciative of comfort. They made many kinds of mats, as well as baskets of tule and sagebrush. They flaked arrowheads and threw them with throwing-sticks (atlatls) to kill game. They cut bone into beads and drilled holes in them with awls made of splintered pieces of bone ground to sharp points. They made fire by rotating wooden drills in wooden sockets (Cressman 1942). They made music on wooden flutes (one artifact flute remains). They played dart games, throwing the darts through hoops, and played gambling games. They made nets and snares of twine and rope twisted from shredded sagebrush bark (indicating that the climate of south central Oregon was desertlike 9,000 years ago). They smoked pipes hollowed out of compacted pumice; one wonders what they smoked in a climate too dry for tobacco, perhaps marijuana.

At the same time, all the highest Stone Age tools were manufactured and used: scrapers, arrowpoints in a range of sizes, all of which could have been thrown with the throwing-stick, and drills and engravers, also manos for grinding and the shallow basin and flat metates on which the manos were used to smash seeds and grains. People also ground colored rocks to make paint for decoration and used pumices for grinding and polishing.

Some of the caves had bits of fired pottery and evidence that humans ate animals now extinct, such as horses and camels (Cressman 1942, 1966). Other animal remains indicate a human diet of birds, namely, pintail, teal, hawk, and sage hen. At Fort Rock, besides now extinct mammals, people ate bison, wolf, fox, and probably bear. Two or three hundred baskets were found, both undecorated and decorated, in the mode of wrapping or embroidery of the weave and by dying and varying the twist of the cords (Cressman 1942).

Cressman (1973) notes that a sandal structurally similar to the Fort Rock Cave sandals but made from different materials has been found at Fish Bone Cave, Nevada, dated at 11,000 years B.P. Probably the shoe industry was ubiquitous because of the comfort and protection it provided for human life. Shoes were therefore made of local material and were decorated according to the whim of the maker. Cressman notes that many artifacts have been given an elegantly decorated finish to delight the beholder, to increase the pleasure of life, a life otherwise filled with fears of killers, for example, the dire wolf, the sabertooth, the jaguar, and fears of the unknown, of the thousands of miles of often frightful landscapes in which people hunted and lived—tundra, taiga, ice-free corridors, bogs, forests, deserts, volcanoes, and lava fields. There was more to their lives than just the struggle to survive, suggests Cressman (ibid.). "There are common solutions which could be arrived at by a reasonably intelligent individual without having to see how some other person perhaps a thousand miles away solved the problem." There was the joy and self-aggrandizement of successful invention. There was the beauty of the artisans' skill, of making highly polished surfaces, beauty of form, beauty of design and color and relief, and beauty of texture, of music and variety. All these things evidently entered into their lives just as in ours.

The next evidence of shoes that we find comes long after in time: in the high civilization of Egypt. Dated 1991 to 1786 B.C. is a painting of Asians bringing rugs and other trading goods to Egypt, depicted on the wall of the tomb of Khnumhotep (fig. 3-7) (Michalowski 1973). The traders wear ankle-high woven cloth boots, probably made of the same tough hemp and wool weave as that in Near Eastern rugs. Whether there is a sack of extra shoes on their donkey

FIG. 3-7. Caravan of Asians wearing shoes, Beni Hasan, rock-cut tomb of Khnumhotep, XII dynasty, 1991 to 1786 B.C. (Michalowski 1973).

FIG. 3-8. King Semenkhara wearing shoes and Princess Meritaten barefoot, XVII dynasty, 1364 to 1361 B.C. (Michalowski 1973).

to replace worn-out shoes we don't know, but their servant, the donkey boy, wears the same kind of shoes as do his masters.

The royalty of Egypt and their gods began to wear shoes in the paintings made soon after. One of the clearest paintings of shoes is that of King Semenkhara and the Princess Meritaten, 1364 to 1361 B.C. (Michalowski 1973), in which the king is wearing beautiful sandals with a thong between his big toe and the other toes, whereas the princess is barefoot (fig. 3-8). This may indicate that women were not to take on the elegance of wearing shoes, but it could hardly indicate that shoes were too expensive for women, seeing that the princess wears the same beautiful beads and diaphanous cottons as does the king.

In our search for substantive evidence of human mitigation of climate and terrain by wearing shoes, we found a Persian terra-cotta

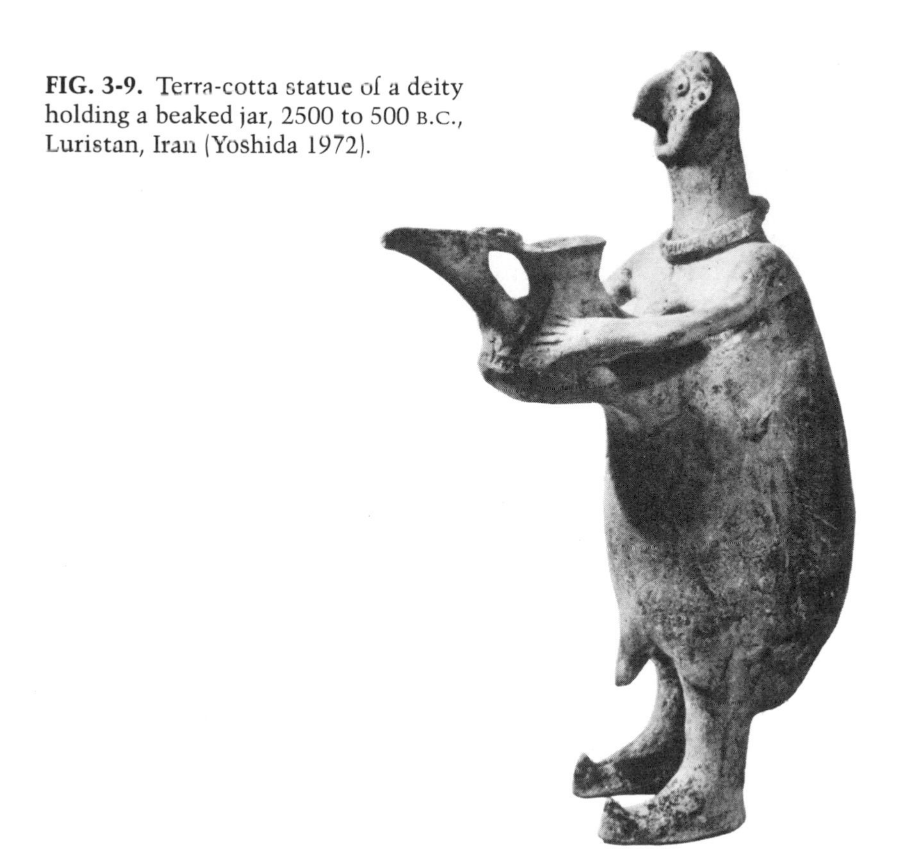

FIG. 3-9. Terra-cotta statue of a deity holding a beaked jar, 2500 to 500 B.C., Luristan, Iran (Yoshida 1972).

FIG. 3-10. Bas-relief libation scene from Malatya, Hittite kingdom of Malid. The king, wearing shoes, pours a libation before the weather god, also wearing shoes, about 1100 B.C.

statue of a deity holding a beaked jar, from Luristan, dated 2500 to 500 B.C. (fig. 3-9) (Yoshida 1972). The great range in estimated time of fabrication is the result of the raiding of Luristan burials by people living near the Zagros Mountains who sold their findings in the European markets and lost the exact provenience of the artifacts. The deity is wearing boots with turned-up toes. Shoes with turned-up toes are still made and sold today in the markets of Iran, Afghanistan, and India.

Our next finding is a bas-relief from the Hittites, 1900 to 1300 B.C., showing the king of Malid (Malatya) wearing shoes, pouring a libation before the weather god, also wearing shoes (fig. 3-10). The artist of Malatya has united in this scene the Anatolian and the Syrian conceptions of the weather god. The concept of such a god seems especially appropriate to our intent of writing about the impact of climate change on human evolution and about human inventions to ameliorate the mischief played by the weather god.

From circa 1000 B.C. have been found ceramic shoes for the coffee table, from Iran, to delight the art lover; they have little if any practical use as cooking pots, for they are hard to clean. The pots are shoes with one pocket for the big toe and a larger pocket for the remaining four toes, much like the socks and boots worn today in Japan (fig. 3-11).

Another coffee-table ceramic, circa 600 B.C., from northwestern

FIG. 3-11. Ceramic pot with human feet wearing shoes, northwest Iran, circa 1000 B.C. Courtesy of Galeria Israel Ltd., Tel Aviv, Israel.

FIG. 3-12. Pottery vessel from Kultepe, Hittite kingdom, 1900 to 1300 B.C. (Gurney 1952).

FIG. 3-13. Brown clay Amlash vessel from northwestern Iran, circa 600 to 500 B.C. Courtesy of S. Dubiner, Tel Aviv, Israel.

FIG. 3-14. Black burnished ware vessel from northern Iran, circa 1350 to 800 B.C. Courtesy of the Los Angeles County Museum of Art, gift of Nasli M. Heeramaneck.

FIGS. 3-15 and 3-16. Amlash ceramic vessels with human feet wearing shoes, northwestern Iran, sixth to fifth century B.C. Courtesy of S. Dubiner, Tel Aviv, Israel.

Iran, is a single ankle-length ceramic boot with a turned-up toe (fig. 3-12). The ceramic itself can have no other purpose but to charm the beholder, because of the uneven heating of its contents were it to be used as cookware. The shoe that it depicts can have no practical purpose, for who ever saw a human, alive or fossil, with toes to fit in the curl? The shoe on which it is modeled must have been woven, or knitted, because the ceramic shows no external stitched seams such as one might have made in constructing the shoe in leather. Other ceramic pots with curled toes are seen in figures 3-13 and 3-14.

Another ceramic of coffee-table fantasy might have been used as a water pitcher (fig. 3-15). Its provenience is sixth to fifth century B.C., northwestern Iran. Its pouring beak is turkeylike. Its water pot is supported on two muscular, solid clay legs shaped like those of a human, with big feet wearing shoes with toes curled twice around, much like a nautilus shell. A deity wearing shoes holds a similar beaked jar (fig. 3-9).

A second very similar water pot of sixth to fifth century B.C., northwestern Iran, with a similar long beak with a turkeylike neck, is decorated with an animal-shaped handle and with coin-shaped medallions around the mouth of the pot (fig. 3-16). Solid human legs

FIG. 3-17. Roman sandal, circa A.D. 100, Vindolanda, Northumberland, England.

hold it up on huge human feet wearing shoes with toes curled twice around, snail shell fashion.

After this time shoes and sandals were undoubtedly well established and ubiquitous. Our next example, chronologically, is a leather sandal found in a Roman encampment along the cross-England Roman wall between England and Scotland, circa A.D. 100, from Vindolanda, Northumberland (fig. 3-17). The sole has two layers of leather nailed together. The engraved leather cover of the top of the foot is held to the sole by a thong passing by the big toe. The high gloss of the leather, even after almost 2,000 years, indicates that shoe polish had been invented.

Roman shoes were about twice as expensive relative to a day's wage than is the case today. We know this from Roland Kent (1920), who translated the maximum prices mandated by the Edict of Diocletian, A.D. 301, into 1920 dollars. He did this by noting that the price of 1 Roman pound of fine gold was set at 50,000 denarii. Changed into the weights and prices in the United States in 1920, the worth of the denarius was 0.434 cents. Using his translation, we may compare the price of Roman shoes with the price of shoes in 1980 and with other modern-day commodities:

	301	1980
Ham per pound	$0.12	$ 5.00
Beef per pound	0.05	2.00
Patrician's shoes	0.65	50.00
Servant's shoes	0.26	20.00
Carpenter's wage per day	0.22	50.00

The modern United States carpenter can pay for shoes with less than one day's work, whereas the Roman carpenter had to work over one day to do so.

At about the same time, during the Han dynasty, the Chinese were wearing elaborate shoes. One of the ceramic figures in the British Museum wears glittering, polished, knee-high leather boots. Another figure wears a suit of chain mail, as well as heavy shoes perhaps clad with metal foil or perhaps brazed of sheet metal. The shoes look much like those worn by helmeted divers today; they are weighted, not articulated as are those of the European armor of the Middle Ages. On the evidence of these figures, the art of shoe crafting in China by the year 2000 B.P. was the equal of today (figs. 3-18 to 3-21).

Somewhat earlier, in Assyria at Babylon, gods and kings alike habitually wore sandals which seem in their representations in stone bas-reliefs to have been made of leather, with soles, heel cups, and toe thongs, with leather thongs to bind the heel cup to the foot crisscrossed over the instep (fig. 3-22) (Paley 1976). For all these various representations of B.C. shoes, there is no hint that stockings had yet been invented.

The role of environment in culture history, according to Cressman (1977), is limiting and permissive. People usually do not have to develop all the opportunities of their environment in order to survive, but the more fully they exploit them the richer becomes their way of life within the limits imposed by the environment. In the case of each aboriginal environment, the limit of exploitation had probably been reached by each indigenous population, in the opinion of Cressman.

The biological variable is the individual. Prehistoric individuals, like modern humans, probably had their greatest period of creativity in youth; but, since their life expectancies were only about twenty years, this period of creativity lasted most of their lives. If the individuals, starting with those of 1 to 3 million years ago in the Olduvai environs, had fully exploited their environment, one expects that they would have invented making pottery, firing pottery, and smelting metal. But they did not, although the food supply and the climate in the Olduvai region were almost as optimal as possible. In fact, for at least a million years of the Pleistocene, *Homo* accomplished little in the way of technological advances. Yet, although there were undoubtedly times of bad climate in the middle latitudes, when deserts took over North Africa and the Near East and when glaciers covered the northern latitudes, there were un-

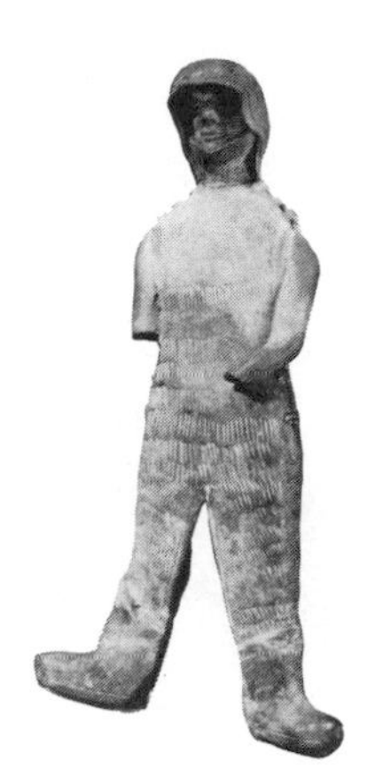

FIG. 3-18. (*left*) Ceramic human figure wearing knee-high, polished, seamed leather boots, Han dynasty, circa 2,000 years B.P., described in the Seligman Collection, catalog by S. H. Hansford and John Ayers, British Museum. Courtesy of Roger Whitfield, assistant keeper. **FIG. 3-19.** (*right*) Ceramic human figure wearing chain mail and heavy boots which are clad with metal foil or perhaps sheet metal, Han dynasty, 206 B.C. to A.D. 220, described in the Seligman Collection, catalog by S. H. Hansford and John Ayers, British Museum. Courtesy of Roger Whitfield, assistant keeper.

FIG. 3-20. (*left*) Life-size terra-cotta infantryman from the Chin dynasty, 221 to 206 B.C. The Metropolitan Museum of Art, on loan from China. **FIG. 3-21.** (*right*) Wood sculpture, late Chou dynasty, 770 to 256 B.C. The Metropolitan Museum of Art, gift of Mathias Komor, 1948.

FIG. 3-22. Stone bas-relief of an Assyrian king with bow, wearing sandals which have a heel, apparently made of leather, leather thongs on the big toes, and leather thongs binding the heel and sole to the foot; filleted genie with pail, wearing sandals like those of the king. Courtesy of the Brooklyn Museum.

doubtedly also times of good climate, with plentiful food and a comfortable way of life outdoors without cave or tent shelters. So one is led to suppose that *Homo* was not yet endowed to invent the technologies that were developed by *Homo sapiens sapiens*. Specifically, early *Homo* has never exploited the environment as fully as today. Now, of course, *Homo sapiens sapiens* is inventing still more exploitive techniques, as witness the release of energy from coal, starting in the last century, and the release of nuclear energy in this century. The biological variable is still the individual; we are each still learning how to dominate the resources of the physical world and the laws of nature and make them serve our comfort and well-being. We are learning to exploit the force of our intelligence.

At an early level of culture, with pressures for survival bearing heavily on an individual and on a group, it is reasonable to suppose that custom was less compelling than later when conditions of life were easier and more organized and class distinctions, the "establishment" and its ways of life, had developed (Cressman 1977). Class structure develops and is associated with abundant food resources, for class differences are a luxury which can be developed and exploited only where there is an abundant food supply and the means and will to exploit it. A wealth of natural resources does not mean that a class-differentiated society must or will develop, but such a society cannot develop without an adequate economic base. The economic base in turn is based on technology acquired and developed over long periods of time in response to human needs for survival among many kinds of climate and climate change and in response to human appreciation of comfort and beauty.

References

Allibone, T. E., Sir Mortimer Wheeler, I. E. S. Edwards, E. T. Hall, and A. E. A. Werner, 1970, *The Impact of the Natural Sciences on Archaeology*, British Academy & Royal Society, Oxford Univ. Press, London.

Barr, Terry N., 1981, Science *214*, 1087–1095.

Bedwell, S. F., 1970, *Prehistory and Environment of the Pluvial Fort Rock Lake Area of South Central Oregon*, Univ. Microfilms, Ann Arbor.

Bedwell, S. F., 1973, *Fort Rock Basin: Prehistory and Environment*, Univ. of Oregon Press, Eugene.

Berger, R., and P. V. Wells, 1967, Science *155*, 1640–1647.

Bishop, W. W., 1971, pp. 493–527 in *Late Cenozoic Glacial Ages*, ed. K. Turekian, Yale Univ. Press, New Haven, Conn.

Braidwood, R. J., 1970, pp. 81–91 in *Radiocarbon Variations and Absolute Chronology*, 12th Nobel Symposium, ed. I. U. Olsson, Almqvist & Wiksell, Stockholm.

Braidwood, R. J., H. Campbell, and W. Schirmer, 1981, J. Field Archaeology *8*, 249–258.

Brain, C. K., 1981, *The Hunters or the Hunted*, Univ. of Chicago Press, Chicago.

Bromehead, C. N., 1954, in *A History of Technology*, ed. C. Singer, E. J. Holmyard, and A. R. Hall, Oxford Univ. Press, London.

Brown, L. R., 1981, Science *214*, 995–1002.
Butzer, K. W., 1964, *Environment and Archaeology*, Aldine, Chicago.
Cervinka, V., W. J. Chancellor, R. I. Cosselt, R. G. Curley, and J. B. Dobie, 1974, Joint Study, California Department of Food and Agriculture and University of California at Davis.
Champion, T. C., 1980, Nature *284*, 513–516.
Clark, G., 1969, *World Prehistory*, Cambridge Univ. Press, London.
Clark, J. D., and H. Kurashina, 1979, Nature *282*, 33–39.
Cope, M. J., and W. G. Chaloner, 1980, Nature *283*, 647–649.
Coppens, Y., 1978, pp. 499–506 in *Geological Background to Fossil Man*, ed. W. W. Bishop, Univ. of Toronto Press, Toronto.
Cressman, L. S., 1942, *Carnegie Inst. Washington Pub. 538*, Washington, D.C.
Cressman, L. S., 1966, American Antiquity *31*, 866.
Cressman, L. S., 1973, *Museum of Natural History Bull. 20*, Univ. of Oregon, Eugene.
Cressman, L. S., 1977, *Prehistory of the Far West: Homes of Vanished Peoples*, Univ. of Utah Press, Salt Lake City.
Emiliani, C., 1978, Earth & Planet. Sci. Lett. *37*, 349–352.
Etkins, R., and E. S. Epstein, 1982, Science *215*, 287–289.
Fairbridge, R. W., 1961, pp. 99–185 in vol. 4 of *Physics and Chemistry of the Earth*, ed. L. H. Ahrens, F. Press, K. Rankama, and S. K. Runcorn, Pergamon Press, New York. See also 1960, Scientific American *202*, 70.
Fitch, F. J., 1972, pp. 77–91 in *Calibration of Hominoid Evolution*, ed. W. W. Bishop and J. A. Miller, Univ. of Toronto Press, Toronto.
Forbes, C. J., 1954, pp. 572–599 in *A History of Technology*, ed. C. Singer, E. J. Holmyard, and A. R. Hall, Oxford Univ. Press, London.
Freed, S. A., and R. S. Freed, 1980, Natural History, Jan., 68–75.
Gates, W. L., 1976, Science *191*, 1638–1644.
Gates, W. L., E. S. Butten, A. B. Kahle, and A. B. Nelson, 1971, *A Documentation of the Mintz Arakawa Two-Level Atmospheric General Circulation Model, R-877*, Rand Corp., Santa Monica, Calif.
Gates, W. L., and M. E. Schlesinger, 1972, Rand Corp., Santa Monica, Calif.
Godwin, H., 1970, Phil. Trans. Roy. Soc. London *269*, 57–75, 1193.
Gowlett, J. A. J., J. W. K. Harris, D. Walton, and B. A. Woods, 1981, Nature *294*.
Green, H. S., C. B. Stringer, S. N. Collcutt, A. P. Currant, J. Huxtable, H. P. Schwarcz, N. Debenham, C. Embleton, P. Bull, T. Molleson, and R. E. Bevins, 1981, Nature *294*, 707–713; same authors in 1981, Antiquity *55*, 184, and in 1982, New Scientist *93*, 21.
Gurney, O. R., 1952, *The Hittites*, Penguin, Bungay, Suffolk, Eng.
Harris, J. W. K., 1981, L. S. B. Leakey Found. News *19*, 1–10.
Harris, J. W. K., and T. Herbich, 1978, pp. 528–549 in *Geological Background to Fossil Man*, ed. W. W. Bishop, Univ. of Toronto Press, Toronto.

Hay, R. L., 1980, Nature *284*, 401–403.
Howell, F. C., 1965, *Early Man*, Life Nature Library, New York.
Howell, F. C., 1966, pp. 88–201 in *Recent Studies in Paleoanthropology*, ed. J. D. Clark and F. C. Howell, American Anthropologist Special Pub. *68*, 88–201.
Howell, F. C., 1972, pp. 311–368 in *Calibration of Hominoid Evolution*, ed. W. W. Bishop and J. A. Miller, Univ. of Toronto Press, Toronto.
Ikawa-Smith, Fumiko, 1980, American Scientist *68*, 134–145.
Isaac, G. L., 1972, pp. 381–430 in *Calibration of Hominoid Evolution*, ed. W. W. Bishop and J. A. Miller, Univ. of Toronto Press, Toronto.
Isaac, G. L., 1978, pp. 139–147 in *Geological Background to Fossil Man*, ed. W. W. Bishop, Univ. of Toronto Press, Toronto.
Johanson, D. C., 1976, National Geographic *150*, 790–811.
Johanson, D. C., 1978, pp. 43–65 in *Science Year Worldbook*, Childcraft International, Inc.
Johanson, D. C., and T. D. White, 1979, Science *202*, 320–321.
Kent, R. G., 1920, Univ. of Pennsylvania Law Rev., 35–47.
Klein, R. G., 1974, Scientific American, June, 96–105.
Leakey, M. D., 1978, pp. 151–170 in *Geological Background to Fossil Man*, ed. W. W. Bishop, Univ. of Toronto Press, Toronto.
Leakey, M. D., 1979, *Olduvai Gorge: My Search for Early Man*, Collins, London.
Leakey, R. L., and R. Lewin, 1978, *The People of the Lake*, Aron/Doubleday, New York.
Libby, W. F., 1952, *Radiocarbon*, Univ. of Chicago Press, Chicago.
Libby, W. F., 1955, *Radiocarbon Dating*, 2d ed., Univ. of Chicago Press, Chicago.
Libby, W. F., 1979, private communications that the possibility of survival of charcoal for a million years is not known.
Los Angeles Herald Examiner, 1980, Mar. 31, p. A3, and Apr. 2, p. A3.
Macintosh, N. W. G., 1972, pp. 7–14 in *8th International Conference on Radiocarbon Dating, Proc.*, ed. T. A. Rafter and T. Grant-Taylor, Lower Hutt, New Zealand.
Madden, R., J. D. Muhly, and T. S. Wheeler, 1977, Scientific American, Oct., 122–131.
Mails, T. E., 1979, *Fools Crow*, Doubleday, Garden City, N.Y.
Martin, P. S., manuscript, Department of Geoscience, University of Arizona, Tucson.
McDougall, Ian, 1981, Nature *294*, 120–124.
Michalowski, K., 1973, *Art of Ancient Egypt*, trans. N. Gaterman, Harry N. Abrams Pub., New York.
Molleson, T., 1977, pp. 77–95 in *British Quaternary Studies*, ed. F. W. Shotten, Clarendon Press, Oxford, Eng.
Munthe, J., 1981, L. S. B. Leakey Found. News *20*, 5–6.

Namias, J., 1980, EOS *61*, 449.
Paley, S. M., 1976, *King of the World: Ashur-nasir-pal II of Assyria 883–859 B.C.*, Brooklyn Museum, New York.
Pilbeam, D., 1982, Nature *295*, 232–237.
Posnansky, M., 1980, Archaeology at UCLA *2*, 4.
Protsch, R., and R. Berger, 1973a, Science *179*, 235.
Protsch, R., and R. Berger, 1973b, Orientalia *42*, fasc. 1–2, 214–227.
Riehl, H., and J. Meitin, 1979, Science *206*, 1178–1179.
Rubin, K., 1981, L. S. B. Leakey Found. News *19*, 9–11.
Schwarzbach, M., 1961, in *Descriptive Paleoclimatology*, ed. A. E. M. Nairn, Interscience Pub., New York.
Schwarzbach, M., 1963, *Climates of the Past*, trans. R. O. Muir, D. Van Nostrand Co., London.
Shackleton, R. M., 1978, pp. 19–28 in *Geological Background to Fossil Man*, ed. W. W. Bishop, Univ. of Toronto Press, Toronto.
Simons, E. L., 1970, pp. 901–922 in *Adventures in Earth History*, ed. P. Cloud, W. H. Freeman & Co., San Francisco.
Singer, C., E. J. Holmyard, and A. R. Hall, 1954, *A History of Technology*, Oxford Univ. Press, London.
Siscoe, G. L., 1979, preprint, submitted to Geophysical Reviews.
Stekelis, M., L. Picard, N. Schulman, and G. Haas, 1960, Bull. Res. Council Israel *9G*, 175–183.
Toffler, A., 1980, New York Times, Mar. 9, pp. 24–30.
Trinkhaus, E., and W. W. Howells, 1979, Scientific American, Dec., 118–133.
Tripathi, V., 1973, in *Radiocarbon and Indian Archaeology*, ed. D. P. Agrawal and A. Ghosh, Tata Inst., Bombay, India.
van der Hammen, T., T. A. Wijmstra, and W. H. Zogwijn, 1971, in *Late Cenozoic Glacial Ages*, ed. K. Turekian, Yale Univ. Press, New Haven, Conn.
Walker, A., and R. E. F. Leakey, 1978, Scientific American, Apr., 54–66.
Wasson, J., 1974, *Meteorites*, Springer-Verlag, Berlin.
Weast, R. C., ed., 1969, *Handbook of Chemistry and Physics*, 50th ed., Chem. Rubber Co., Cleveland.
Wells, P. S., B. Benefit, C. C. Quillian, and J. D. Stubbes, Jr., 1981, J. Field Archaeology *8*, 289–302.
Wendorf, F., R. Schild, and R. Said, 1970, pp. 57–79 in *Radiocarbon Variations and Absolute Chronology*, 12th Nobel Symposium, ed. I. U. Olsson, Almqvist & Wiksell, Stockholm.
White, T. D., 1980, Science *208*, 175–176.
Woillard, G. M., and W. G. Mook, 1982, Science *215*, 159–161.
Yoshida, M., 1972, *In Search of Persian Pottery*, Weatherhill/Tankosha, New York/Tokyo.

Appendix 1. Equations Explanatory of the Text

For the introduction, page 9, the calculation of the ratio of vapor pressure of H_2 to HD,

$$\ln_{(P(H_2)/P(HD))} = \frac{E_0(H_2) - E_0(HD)}{RT} + \frac{F_s(H_2) - F_s(HD)}{RT} + \frac{3}{2} \ln \frac{M(H_2)}{M(HD)} \qquad (1)$$

where the free energies F_s are computed at temperatures T from the Debye theory of solid state, E_0 is the zero point energy of the solid due to Debye vibrations, and M is the molecular mass.

For table 2-2, changes in natural $^{13}C/^{12}C$, we write the enrichment E at time t for ^{13}C as

$$E_t = 0.027\,(M/M^*)\,(1 - 0.5\times10^{-3}\,(T - T^*))\,(1 + 8.5\times10^{-3}\,(T - T^*)) \qquad (2)$$

where the number 8.5×10^{-3} is assumed for the average fraction of ^{13}C at the sea-air interface and where T is the average temperature in degrees C at time t, T^* is the average temperature today, assumed to be 20° C, M is the mass of the biosphere at time t, and M^* is the mass today.

Appendix 2. The Slope of Eight

The linear relationship of the isotope ratios of hydrogen and oxygen in rain and snow was discovered by H. Craig (1961) to be

$$\delta_D = 8\delta_{18} + C \tag{1}$$

where C is a very small constant,

$$\delta_D \equiv \left(\frac{(D/H)}{(D/H)_{SMOW}} - 1 \right) \times 10^3 \tag{2}$$

$$\delta_{18} \equiv \left(\frac{(^{18}O/^{16}O)}{(^{18}O/^{16}O)_{SMOW}} - 1 \right) \times 10^3 \tag{3}$$

and where the ratios labeled SMOW are those for standard mean ocean water.

This relationship has been abundantly verified for rain and snow collected worldwide and measured by the IAEA since 1969 (IAEA 1969–1975). The IAEA data are shown in figure 1-1 as a plot of δ_D and δ_{18} with the characteristic slope of 8.

This slope can readily be computed from laboratory measurements (Friedman 1974; Stewart and Friedman 1975); for the temperature range of 0° to 35° C these measurements can be represented as

$$\frac{(D/H)_{liq}}{(D/H)_v} = (-9\times10^{-4})T + 1.1035 \tag{4}$$

$$\frac{(^{18}O/^{16}O)_{liq}}{(^{18}O/^{16}O)_v} = (-9\times10^{-5})T + 1.01135 \tag{5}$$

where the subscripts *liq* and *v* mean liquid and vapor and where T is the temperature in degrees C. We take the ratio for the liquid to be equal to the ratios for ocean water (SMOW).

So, for rain (made from vapor distilled from the ocean):

$$\delta_D = \left(\frac{(D/H)_v}{(D/H)_{liq}} - 1 \right) \times 10^3 \tag{6}$$

$$= \frac{10^3}{(-9 \times 10^{-4})T + 1.1035} - 10^3$$

$$\delta_{18} = \left(\frac{(^{18}O/^{16}O)_v}{(^{18}O/^{16}O)_{liq}} - 1 \right) \times 10^3 \tag{7}$$

$$= \frac{10^3}{(-9 \times 10^{-5})T + 1.01135} - 10^3$$

Thus the derivatives of δ_D and δ_{18} with respect to temperature are

$$\frac{\Delta(\delta_D)}{\Delta T} = \frac{0.9}{((-9 \times 10^{-4})T + 1.1035)^2} \tag{8}$$

$$\frac{\Delta(\delta_{18})}{\Delta T} = \frac{0.09}{((-9 \times 10^{-5})T + 1.01135)^2} \tag{9}$$

The ratio of these two derivatives gives the slope of 8 of the precipitation curve in figure 1-1. In fact, the IAEA data yield a far more accurate determination of the slope than can be made in the laboratory.

From these laboratory relationships, we compute the values of δ_D and δ_{18} for rain made by distillation from the ocean at 20° C as −80 and −10 ppt respectively. Thus, when rain and snow show very large depletions, for example, δ_D of −200 to −300 ppt, two or three distillations have occurred. It is known from measurements of tritium in rain (Levinthal and Libby 1968) that two or three distillations occur in the U.S. between evaporation from the Pacific Ocean in the west and precipitation on the east coast. The agreement between the number of distillations deduced from stable isotopes and from tritium is remarkable.

For every point on the curve in figure 1-1 characterizing rain and snow, there is a corresponding average air temperature, as may be seen in figures 1-2 and 1-3. Consequently, we looked for a similar dependence and the slope of 8 to occur in trees which incorporate rainwater. We found the slope of 8 in a Japanese cedar (Libby et al. 1976) measured for the years 160 to 1900. But, in the sequence of German oaks which we measured (Libby and Pandolfi 1973, 1974), we reported that we did not find a slope of 8, and we listed our measurements.

We now state that a slope of 8 obtains in all the trees which we have measured, including the oaks. The explanation is as follows. For the oaks, we measured all isotope ratios, using wood for the years 1712 to 1714 as our standard. In those years Europe was very cold, near the bottom of the First Little Ice Age according to our tree measurements (see fig. 2-2) and according to European thermometer records. The wood we used as our standard was consequently greatly depleted in heavy isotopes, for example, in deuterium by −180 ppt, because the rain incorporated was correspondingly depleted.

When our measured, listed, and published isotope measurements, referred to wood of 1712 to 1714, are referred instead to ocean water (SMOW), then $\delta_{D_{SMOW}}$ and $\delta_{18_{SMOW}}$ are related to each other by the slope of 8 for the oak sequence, just as they are in the cedar sequence (see fig. 1 of this appendix).

The relationship between the TREE δ_D and SMOW δ_D is (in ppt)

$$\delta_{D_{SMOW}} = 0.82\delta_{D_{TREE}} - 180 \tag{10}$$

which follows from the relationships, for deuterium:

$$\delta_{D_{SMOW}} = \left(\frac{R_{TREE}}{R^*_{SMOW}} - 1 \right) \times 10^3 \tag{11}$$

$$\delta_{D_{TREE}} = \left(\frac{R_{TREE}}{R^*_{TREE}} - 1 \right) \times 10^3 \tag{12}$$

Therefore,

$$R^*_{SMOW}\,(\delta_{D_{SMOW}} \times 10^{-3} + 1) = R^*_{TREE}\,(\delta_{D_{TREE}} \times 10^{-3} + 1) \tag{13}$$

so that for 1712 to 1714, when $\delta_{D_{TREE}} = 0$ and $\delta_{D_{SMOW}} = -180$:

$$\frac{R^*_{TREE}}{R^*_{SMOW}} = 1 - 0.18 = 0.82 \tag{14}$$

Finally, we now predict that, for all trees which subsist on rainwater, the deuterium and oxygen isotope ratios will be related linearly in their rings with a slope of 8, namely:

$$D/H = 8(^{18}O/^{16}O) + \text{constant} \tag{15}$$

The constant is different in the oaks and in the cedar from its nearly zero value in seawater. In plotting figure 1 of this appendix we subtracted the appropriate constant from the tree measurements in order to make all the measured points lie on the seawater curve; this subtraction has no effect on the slope of 8, of course.

We hope that analysis of stable isotopes in tree ring sequences will become an international undertaking; that, with the collaboration of many laboratories in many countries, the history of the climate in recent times, perhaps as far back as 30,000 years, will be deduced; and that, from this history, the likely trends of climate in the near future will be predicted.

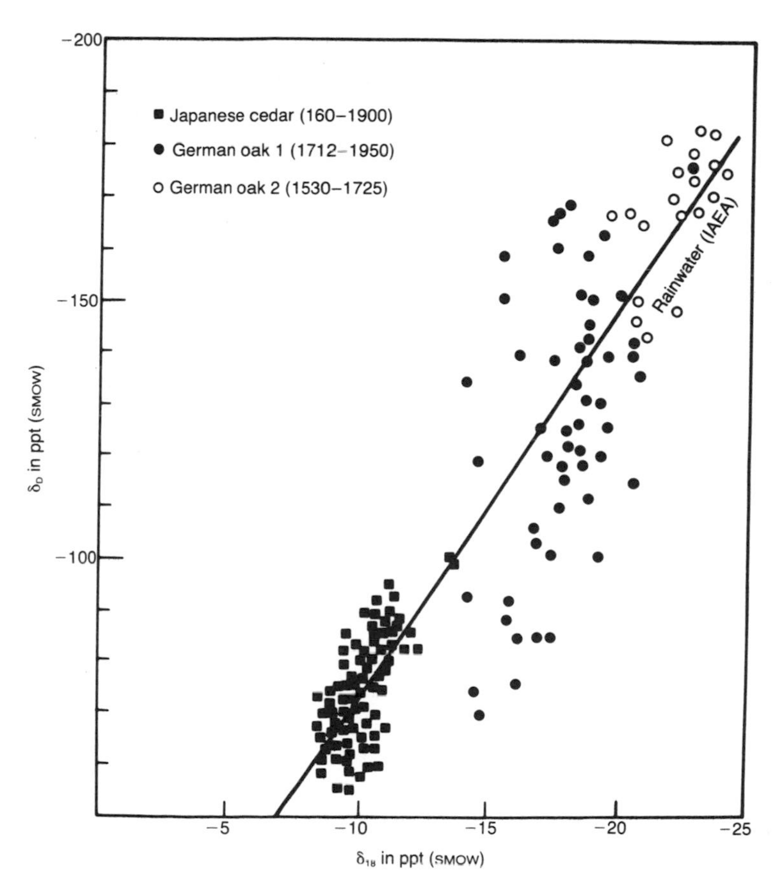

FIG. A-1. Deuterium isotope ratio versus oxygen isotope ratio measured in our sequence of German oaks and in our Japanese cedar. All ratios are related to each other by the slope of 8, namely, according to $\delta_D = 8\delta_{18} +$ constant. We have subtracted the constant in order to make all points lie on the line for worldwide precipitation (see fig. 1-1).

References

Craig, H., 1961, Science *133*, 1702–1703.

Friedman, I., 1974, CLIMAP Conference, Oct., University of Miami, Coral Gables.

IAEA (International Atomic Energy Agency), 1969–1975, *Environmental Isotope Data, Nos. 1–5*, Vienna.

Levinthal, J. S., and W. F. Libby, 1968, J. Geophys. Res. *73*, 2715–2719.

Libby, L. M., and L. J. Pandolfi, 1973, Contribution 219, Colloques Internationaux du Centre National de la Recherche Scientifique, Gif-sur-Yvette, France.

Libby, L. M., and L. J. Pandolfi, 1974, Proc. Nat. Acad. Sci. *71*, 2482–2486.

Libby, L. M., L. J. Pandolfi, P. N. Payton, J. Marshall III, B. Becker, and V. G. Siebenlist, 1976, Nature *261*, 284–288.

Stewart, M. K., and I. Friedman, 1975, J. Geophys. Res. *80*, 3812–3818, and references therein.

Appendix 3. The Theory of Isotope Fractionation in Cellulose

The computation of isotope fractionation systematized by Urey (1947) from the work of several other people is readily applied to compute such fractionations in the formation of cellulose in plants. It was basic to do this in order to assess the magnitude of temperature effects to be expected and to make sure they would be small compared to fractionation in precipitation, before beginning isotope measurements in wood (Libby 1972).

Isotope fractionation at equilibrium can be computed from the usual thermodynamic relations between the free-energy change in a chemical reaction and the equilibrium constant for the reaction according to

$$\Delta G = -RT \ln K \tag{1}$$

where ΔG is the free-energy change at unit pressure and K is the equilibrium constant. For the reaction

$$\mathrm{AH} + \mathrm{BD} \leftrightarrow \mathrm{BH} + \mathrm{AD}$$

K depends on the concentrations of the reactants according to

$$K = \frac{[\mathrm{BH}][\mathrm{AD}]}{[\mathrm{AH}][\mathrm{BD}]} \tag{2}$$

See, for example, Mayer and Mayer 1940. T is the absolute temperature in degrees Kelvin and R is the gas constant, 2 calories per gram degree.

Equation 1 states that a reaction tends to proceed in the direction in which energy is emitted. The equilibrium is displaced in that direction, but, depending on the magnitude of ΔG, the components on the side of energy emission will be present in greater concentrations at equilibrium. K, defined in equation 2, is the reaction rate in the forward direction divided by that in the reverse direction.

The change in free energy in the reaction depends on the change in vibrational energies, ω, between AH and AD and between BH and BD. The force constant k depends on the electron configuration and hence to a good approximation is the same in AH and AD, but the vibrational frequencies ω depend both on k and on reduced mass M and therefore are different, according to

$$\omega_{AH} = \left(\frac{k}{M_{AH}}\right)^{1/2} \quad (3a)$$

$$\omega_{AD} = \left(\frac{k}{M_{AD}}\right)^{1/2} \quad (3b)$$

and similarly for BH and BD:

$$\omega_{BH} = \left(\frac{k'}{M_{BH}}\right)^{1/2} \quad (4a)$$

$$\omega_{BD} = \left(\frac{k'}{M_{BD}}\right)^{1/2} \quad (4b)$$

where k and k' are not equal to each other.

It follows from Urey 1947 that

$$K = e^{(-h/2\pi)(\omega_{AH} - \omega_{AD})/kT} e^{(-h/2\pi)(\omega_{BD} - \omega_{BH})/kT} \quad (5)$$

namely, that the isotope fractionation at equilibrium depends only on the temperature and the reduced masses of the reacting molecules. This is why the method of tree thermometers by measuring isotope ratios in wood is truly a thermometer, independent of the amount of water and nutrients that the tree receives.

K is usually written, for example, as

$$K = \frac{Q_{AH} Q_{BD}}{Q_{AD} Q_{BH}} \quad (6)$$

where the Q's are called partition functions:

$$Q = Q_{translation} Q_{vibration} Q_{rotation} \quad (7)$$

For liquids and solids, $Q_{translation} = 1$, and the energy barriers preventing free internal rotation are so high at temperatures around 300° K that rotations deteriorate into vibrations. In this case, Q reduces to

$$Q = Q_{vib} = \sum_i \sum_j g_i e^{(-\omega_i (h/2\pi) kT)} e^{(-\Delta E_j / kT)} \quad (8)$$

$j = 0, 1, 2, \ldots$

Table A-1. Stretching Vibrations Assumed for Cellulose

Bond	Stretching Vibration, Cm.$^{-1}$
| —C—H |	2960
| | —C—C— | |	900
—O—H	3680
| —C—(OH) |	(1200)†
^{16}O—^{16}O	1580
	B = 1.44560

†Calculated from a force constant taken as ½ of force constant for $C = 0$ (Herzberg 1945: 195, tables 51, 89).

Table A-2. Bending Vibrations Assumed for Cellulose

Bending Vibration	Bending Frequency, Cm.$^{-1}$
H ↙↗ | R—C—OH | R	920
H | ↖↘ R—C—OH | R	920
H | R—C—OH | ↙↗ R	920
H | ↔ R—C—O—H | R	(700)†
H | R—C—OH ↖↘ | R	(375)‡

†Herzberg 1945: 196, tables 51, 89.
‡Herzberg 1945: tables 118, 119.

which can be written

$$Q_{vib(T)} = \Pi_j g_i \frac{e^{-\Delta E_i/kT}}{(1 - e^{(-h/2\pi)\omega_i/kT})} \tag{9}$$

where k is the Boltzman constant, T is the absolute temperature, ω_i is the frequency of the ith appropriate vibration of degeneracy g_i, j is the vibrational quantum number, and ΔE_i is the difference in zero point energy between the isotope variations of the molecules. In the application to isotope fractionation in the formation of cellulose, because the energy barriers preventing free internal rotation are high so that rotations degenerate into torsional oscillations and bending vibrations, these will be the main dependencies of Q_{vib}.

The stretching and bending vibrations appropriate to cellulose, which is a chain molecule of module (H-C-OH), are listed in tables A-1 and A-2. Their frequencies differ slightly for molecules containing different isotopes, as shown in equations 3 and 4, for example. Substituting in equation 9 the frequencies in table A-1 and noting that the torsional frequency, 920 cm.$^{-1}$, is threefold degenerate, one computes $Q(273° \text{ K}) = 1.2302$ and $Q(298° \text{ K}) = 1.2981$ for the module (H-C-OH).

In principle the D-H isotope effect in cellulose involves two possible substitutions, namely, substitution of a deuteron for the hydrogen linked to the carbon atom and substitution of the hydrogen linked to the carbon. In the case that groups of atoms designated R and R' are attached to the pivotal carbon atom, the appropriate reduced mass is insensitive to substitution of H by D because the masses of R and R' are almost infinitely large relative to H or D. The partition functions computed for the various isotopic substitutions of the cellulose module are given in table A-3.

Table A-3. Partition Functions Computed for the Module of Cellulose (H-C-OH) and Its Isotopic Modifications

Partition Function	273° K	298° K
$Q_{(HCOH)}$	1.2302	1.2981
$Q^*_{(DCOH)}$	40.850	32.695
$Q^*_{(HCOD)}$	35.014	28.086
$Q^*_{(^{13}CH_2O)}$	1.4240	1.4844
$Q^*_{(CH_2{}^{18}O)}$	1.5620	1.6659

Because of the crudeness inherent in assuming bond frequencies for cellulose to be equal to those measured in rather small organic molecules, we neglect anharmonic corrections and complications caused by hydration in solution and effect of hydrogen bond formation.

The fractionation ratios can be written as follows. For the oxygen isotopes, there are three equilibriums corresponding to isotopic exchange between cellulose and each of the three oxygen-containing molecules involved in the reaction:

$$K_{(^{18}O_2)} \equiv \frac{Q^*_{(CH_2{}^{18}O)} Q_{(^{16}O_2)}}{Q_{(CH_2{}^{16}O)} Q^*_{(^{16}O^{18}O)}} \quad (10a)$$

$$K_{(C^{18}O_2)} \equiv \frac{Q^*_{(CH_2{}^{18}O)} Q_{(C^{16}O_2)}}{Q_{(CH_2{}^{16}O)} Q^*_{(C^{16}O^{18}O)}} \quad (10b)$$

$$K_{(H_2{}^{18}O)} \equiv \frac{Q^*_{(CH_2{}^{18}O)} Q_{(H_2{}^{16}O)}}{Q_{(CH_2{}^{16}O)} Q^*_{(H_2{}^{18}O)}} \quad (10c)$$

where Q^* is the partition function for the heavier isotope.

For the hydrogen isotopes two equilibriums enter, so that

$$K_{(HCOD)} \equiv \frac{Q^*_{(HCOD)} Q_{(H_2O)}}{Q_{(CH_2O)} Q^*_{(HDO)}} \quad (11a)$$

$$K_{(DCOH)} \equiv \frac{Q^*_{(DCOH)} Q_{(H_2O)}}{Q_{(CH_2O)} Q^*_{(HDO)}} \quad (11b)$$

and for the carbon isotopes only one equilibrium enters, that with CO_2, so that

$$K_{(^{13}C)} = \frac{Q^*_{(^{13}CH_2O)} Q_{(^{12}CO_2)}}{Q_{(^{12}CH_2O)} Q^*_{(^{13}CO_2)}} \quad (12)$$

But what the chemist actually measures (Urey 1947) is an isotopic fractionation a. For example, for oxygen the fractionation factor corresponding to equation 10a is

$$\frac{1}{a} = \frac{(2[^{18}O_2] + [^{18}O^{16}O])([CH_2{}^{16}O])}{([^{18}O^{16}O] + 2[^{16}O_2])([CH_2{}^{18}O])} \quad (13)$$

Considering the reaction

$$^{18}O_2 + {}^{16}O_2 \leftrightarrow 2(^{18}O^{16}O)$$

the equilibrium constant is 4, neglecting a tiny correction for isotope effect (ibid.):

$$\frac{([^{18}O^{16}O])^2}{[^{18}O][^{16}O]} = 4 \tag{14}$$

because $^{18}O_2$ and $^{16}O_2$, being symmetrical, have only half as many rotational states as the asymmetrical molecule $^{16}O^{18}O$. Substituting 14 into 13, the fractionation factor reduces to

$$a_{(^{18}O_2)} = \frac{Q_{(^{16}O_2)}{}^{1/2} Q^*_{(CH_2{}^{18}O)}}{Q^*_{(^{18}O)}{}^{1/2} Q_{(CH_2{}^{16}O)}} \tag{15}$$

Corresponding, $C^{18}O_2$ and $C^{16}O_2$, being symmetric, have only half the rotational states as does $C^{18}O^{16}O$, so that

$$a_{(C^{18}O_2)} = \frac{Q_{(C^{16}O)}{}^{1/2} Q^*_{(CH_2{}^{18}O)}}{Q^*_{(C^{18}O_2)}{}^{1/2} Q_{(CH_2{}^{16}O)}} \tag{16}$$

but for $a_{(HCOD)}$ and $a_{(DCOH)}$ the isotope effect is large, so that

$$a_{(HCOD)} = \frac{Q^*_{(HCOD)}}{Q_{(CH_2O)}} \left(\frac{2Q_{(H_2O)} + Q_{(HDO)}}{Q_{(HDO)} + 2Q_{(D_2O)}} \right) \tag{17a}$$

$$a_{(DCOH)} = \frac{Q^*_{(DCOH)}}{Q_{(CH_2O)}} \left(\frac{2Q_{(H_2O)} + Q_{(HDO)}}{Q_{(HDO)} + 2Q_{(D_2O)}} \right) \tag{17b}$$

Here, while H_2O and D_2O are symmetric, none of the deuterated or hydrogenated cellulose modules is. Finally, because $^{13}CO_2$ and $^{12}CO_2$ have the same number of rotational states, also $H_2{}^{16}O$ and $H_2{}^{18}O$, the fractionation factors are

$$a_{(H_2{}^{18}O)} = \frac{Q_{(H_2{}^{16}O)} Q^*_{(CH_2{}^{18}O)}}{Q^*_{(H_2{}^{18}O)} Q_{(CH_2{}^{16}O)}} \tag{18}$$

$$a_{(^{13}CO_2)} = \frac{Q_{(^{12}CO_2)} Q^*_{(^{13}CH_2O)}}{Q^*_{(^{13}CO_2)} Q_{(^{12}CH_2O)}} \tag{19}$$

For the isotope ratios D/H, $^{13}C/^{12}C$, and $^{18}O/^{16}O$, we can now compute the partition function ratios $(Q^*/Q)_{CH_2O}$. The partition function ratios for $(Q^*/Q)_{oxygen}$ are taken from Urey (1947), those for $(H_2O)_{liquid}$ are taken from computations of Bottinga (1967), and those for $(CO_2)_{gas}$ are from Bottinga (1967) and for HDO and D_2O from Urey (1947).

The ratios appropriate for each isotope exchange are listed in tables A-4 to A-6. By substituting these in equations 15 through 19,

Table A-4. Partition Function Ratios Computed for Deuterated Cellulose

Partition Function	273° K	298° K
$Q^*_{(DCOH)}/Q\dagger_{(HCOH)}$	33.206	25.187
$Q^*_{(HCOD)}/Q_{(HCOH)}$	28.462	21.636
$(Q^*_{(D_2O)_{gas}}/Q_{(H_2O)_{gas}})^{1/2}$‡	16.503	12.543
$Q^*_{(HDO)_{gas}}/Q_{(H_2O)_{gas}}$	32.7400	24.9460

Note: The partition function ratio for liquid is obtained by multiplying the partition function ratio for gas by the D/H ratio in liquid divided by that for gas. It is taken here as 1.104 at 273° K (Merlivat, Botter, and Nief 1963).
† Table A-3.
‡ Urey 1947.

Table A-5. Partition Function Ratios Computed for Cellulose Containing Heavy Carbon-13

Partition Function	273° K	298° K
$Q^*_{(^{13}CH_2O)}/Q_{(^{12}CH_2O)}$†	1.1575	1.1435
$\ln Q^*_{(^{13}CO_2)}/Q_{(^{12}CO_2)_{gas}}$‡	0.19732	0.17558
$Q^*_{(^{13}CO_2)}/Q_{(^{12}CO_2)_{gas}}$	1.2181	1.1919

† Table A-3.
‡ Bottinga 1967.

Table A-6. Partition Function Ratios Computed for Cellulose Containing Heavy Oxygen

Partition Function	273° K	298° K	Source
$Q^*_{(CH_2{}^{18}O)}/Q_{(CH_2{}^{16}O)}$	1.2697	1.2833	Libby 1972
$\frac{1}{2}(Q^*_{(^{18}O_2)}/Q_{(^{16}O_2)})_{gas}$	1.0923	1.0818	Urey 1947
$\frac{1}{2}\ln(Q^*_{(C^{18}O_2)_{gas}}/Q_{(C^{16}O_2)_{gas}})$	0.12530	0.11108	Bottinga 1967
$\frac{1}{2}(Q^*_{(C^{18}O_2)}/Q_{(C^{16}O_2)_{gas}})$	1.1336	1.1175	Libby 1972
$\ln(Q^*_{(H_2{}^{18}O)_{gas}}/Q_{(H_2{}^{16}O)_{gas}})$	0.06822	0.06164	Bottinga 1967†
$Q^*_{(H_2{}^{18}O)}/Q_{(H_2{}^{16}O)_{gas}}$	1.0706	1.0635	Libby 1972

† The partition function ratio for liquid water is obtained by multiplying the partition function ratio for water vapor with the $^{18}O/^{16}O$ ratio in liquid water divided by the ratio in the vapor. This correction is taken as 1.01150 at 273° K and as 1.00930 at 298° K (Bottinga 1967: 806).

the fractionation factors have been calculated for 273° K and 298° K. The corresponding temperature coefficients are listed in table A-7.

The temperature coefficient so calculated for $^{13}C/^{12}C$ of 0.36 percent per degrees C agrees with the coefficient measured in $^{13}C/^{12}C$ in plant and animal material by Sackett and his coworkers (Sackett et al. 1965; Degans et al. 1968), so that the assumption of equilibrium appears to be more or less valid, and one can hope that the like calculations for oxygen and hydrogen may be meaningful.

At equilibrium, the temperature coefficient of HDO/H_2O is 312 percent (Urey 1947), much larger than that of the D/H ratio calculated for cellulose, so that measuring it in a geographical distribution of the same kind of tree would measure the distribution of D/H in rainwater, and measuring it in a chronological sequence would measure the time dependence of rainwater in the past.

It is important that CO_2 mixes rapidly through the entire global atmosphere in a time of about 5 years and with the oceans (Bien and Suess 1967; Suess 1970; Craig 1957) in about 15 years, so that at any time the atmospheric CO_2 isotope ratios for oxygen reflect the temperature dependent isotope separation in the sea surface. Equilibration between CO_2 gas and seawater is known to occur in less than 48 hours (Li and Tsue 1971). So whether oxygen in wood derives from rain distilled off the sea surface or from CO_2 is not meaningful because they are in equilibrium.

Table A-7. Temperature Coefficients Computed for Isotopes of C, H, and O in Cellulose

Fractionation	273° K	298° K	Temperature Coefficient, Ppt per Degree C
$a_{(^{18}O_2)}$	1.1624	1.1863	+0.96
$a_{(C^{18}O_2)}$	1.1200	1.1484	+1.14
$a_{(H_2{}^{18}O)}$	1.1725	1.1956	+0.92
$a_{(^{13}CO_2)}$†	0.9503	0.9594	+0.36
$a_{(HCOD)}$	1.5881	1.5983	+0.4
$a_{(DCOH)}$	1.8095	1.8606	+2.0

Note: The computed temperature coefficients listed above agree very well with the value 0.35 ppt per degree C measured for $^{13}C/^{12}C$ by Degans et al. (1968) and with the value 3 ppt per degree C measured by Schiegl (1972).

†See equation 19.

In the formation of cellulose from atmospheric CO_2 and rainwater, schematically according to

$$CO_2 + H_2O \rightarrow \text{cellulose} + \text{evolved } O_2$$

all three isotope ratios, D/H, $^{18}O/^{16}O$, and $^{13}C/^{12}C$, are independent thermometers. Whereas in water the oxygen ratio is functionally related to the hydrogen ratio because they are tied through the same bond, they are independent in cellulose because diatomic oxygen is being evolved, and so rotational and vibrational energies of the O-O bond are involved, and because in cellulose oxygen is bound to carbon as well as to hydrogen. For the same reason, the temperature coefficients of $^{18}O/^{16}O$ and $^{13}C/^{12}C$ are independent. Finally, the ratios $^{13}C/^{12}C$ and D/H are independent because carbon and hydrogen are initially in different molecules.

The temperature dependence R of the isotope ratios can be expressed as

$$R_2 = aT + b \qquad (20a)$$

$$R_{13} = cT + d \qquad (20b)$$

$$R_{18} = eT + f \qquad (20c)$$

where a, c, and e are the temperature coefficients and T is the temperature. Then the change in temperature with time, determined by measuring the changes in the ratios in a tree ring sequence over a span of years or centuries, ΔT, is given by

$$\Delta T = R_2/a = R_{13}/c = R_{18}/e \qquad (21)$$

The multiple overdetermination of temperature change by using a set of several thermometers makes it possible to show that a temperature change did occur.

One may ask whether there is isotope exchange between the hydrogens bound in old interior cellulose and in new sap. As table A-1 shows, both hydrogens are bound tightly, even tighter than C-C bonds, so that, knowing that radiocarbon dates in tree rings mainly agree with the dates obtained by counting the rings, meaning that in heartwood carbon does not exchange, we expect the same to be true of hydrogen and oxygen in heartwood. This is true as shown by experiments with dyes ingested by trees which remain in the ring of the year of ingestion and as shown by injection of radiocarbon sugar solutions which stay where they are injected and do not diffuse.

Even old tree ring sequences which are not tied to the present by overlapping sequences of trees can be radiocarbon-dated and analyzed for isotope variations. In this way temperature changes of the past, as far back as trees remain from them, can be determined.

As to the question of whether the proportion of lignin and cellulose in wood changes from year to year, we have analyzed the bond strengths and abundances in lignin compared to wood and find that as much as a 10 percent variation can affect the temperature coefficients of the isotope ratios by only about 1.5 percent; we feel that this uncertainty is tolerable in determining the variation of temperatures in climates of the past.

We have shown that historic climate changes have produced changes in stable isotope ratios in tree rings by as much as 6 parts in 20, that is, by 30 percent. Hence errors of ~1.5 percent are acceptable at present but may require understanding and avoiding when the method of tree thermometry becomes more refined.

References

The references for appendix 3 are included in the list of references for the introduction. See especially Urey 1947.

Name Index